AN INTRODUCTION TO STRESS AND HEALTH

AN INTRODUCTION TO STRESS AND HEALTH

Edited by

Dr. Parikshit Shirode

Delve PUBLISHING

www.delvepublishing.com

An Introduction to Stress and Health

Dr. Parikshit Shirode

Delve Publishing
4164 Lakeshore Road
Burlington, ON L7L 1A4
Canada
www.delvepublishing.com
Email: orders@arclereducation.com

ISBN: 978-1-77956-392-7 (Hardcover)

ABOUT THE EDITOR

Dr Parikshit Shirode, an eminent surgeon from Ayurveda fraternity, completed his graduation as BAMS with flying colours followed by M.S. in Shalya Tantra (General Surgery) from V. A. M. V., Amaravati, Maharashtra holding 2nd position at the university level. He has profound clinical experience of 21 years. He has managed various difficult cases using his expertise in para-surgical and surgical skills as well as with conventional Ayurvedic medicine. Almost 2200+ cases have been operated successfully by him. He has 16 years of UG Teaching and 7 years of PG teaching experience. He started his academic career as a lecturer in Shalya Tantra at Shree Saptashrungi Ayurveda Mahavidyalaya and hospital, Nashik, Maharashtra, and continued there as an associate professor. In December 2018, he joined Parul Institute of Ayurved, Vadodara, Gujarat as a Professor at the department of Shalyatantra. Till now 15 PG Scholars have submitted their dissertations successfully under his keen guidance, and at present 9 PG Scholars are doing their dissertations work under his guidance. He has extended his professional skills towards social cause by organizing and participating in various diagnostics and treatment camps. Participation in 'Hemalkasa Surgical Camp at Lokbiradari Prakalp, Hemalkasa, Dist. Gadchiroli, Maharashtra (A project by Dr Prakashji Amte and Rotary, Nagpur) in 2005, 2006 and 2007 as a volunteer surgeon where he was awarded with the certificate of excellence by Rotary Nagpur. He has also received prestigious awards like Sushruta award by Gujarat board of Ayurveda and Unani system of medicine, Vikram Sarabhai Innovation Award by the hands of Honorable president of NCISM, Dr B. S. Prasad and Piyush Mahodadhi Award by NIMA Surgical Society. He has published 25+ articles in renowned international journals. Articles authorized by him are also published in proceedings of various national and international webinars and seminars. He also contributed his expertise through writing the chapters in books published by international publishers. His innovative idea 'Modified Multi Needle Therapeutic Blood Lancing Device' has received Device Design patent by the Patent Authority of India.

TABLE OF CONTENTS

LIST OF FIGURES

ABSTRACT

This book offers a thorough examination of the intricate connection between stress and health, delving into the ways in which stress affects both the body and the mind. In this comprehensive analysis, the authors explore the multitude of stressors that individuals encounter, including work-related pressures, interpersonal dynamics, and environmental influences. They thoroughly investigate strategies for effectively navigating and handling stress. In addition, the book delves into the physiological reactions to stress, specifically examining the involvement of the nervous and endocrine systems in the stress response. This book, "An Introduction to Stress and Health," provides valuable insights into the intricate relationship between stress and health. It is highly recommended for individuals who are interested in gaining a deeper understanding of their well-being and how to enhance it.

PREFACE

The correlation between stress and health has become a subject of growing interest in recent years. In our journey through the intricacies of contemporary existence, it is vital to comprehend the effects of stress on our overall health and wellness.

The significance of this book resides in its capacity to illuminate the crucial role that stress plays in our lives and its profound influence on our health. Stress has become a widespread problem that affects people of all ages and backgrounds in today's fast-paced society. Gaining insight into the mechanisms by which stress impacts our health is essential for developing efficacious interventions and strategies to alleviate its adverse effects.

The information contained in this book is urgently needed due to the escalating global prevalence of stress-related health issues. Chronic stress has been associated with a wide range of health issues, such as cardiovascular diseases, mental health disorders, and immune system dysfunction. This book offers a thorough examination of stress and its impact on health, equipping readers with the necessary knowledge and tools to effectively address and manage stress.

This book will attract a diverse audience, encompassing students, researchers, healthcare professionals, and individuals with a keen interest in comprehending the intricate correlation between stress and health. This book is a valuable resource for students studying psychology, public health, medicine, or related fields. This book offers a comprehensive examination and valuable perspectives that will be advantageous for researchers seeking to broaden their understanding of the relationship between stress and health.

Readers will acquire a more profound comprehension of the complex mechanisms by which stress affects our health by immersing themselves in the pages of "An Introduction to Stress and Health." The users will acquire knowledge regarding the physiological and psychological impacts of stress, the significance of stress in different health conditions, and efficient techniques for dealing with and controlling stress. With this knowledge, readers will be more capable of identifying, dealing with, and reducing the adverse impacts of stress on their physical and mental health.

Ultimately, "An Introduction to Stress and Health" is a thorough and perceptive book that provides a multifaceted view of the intricate connection between stress and health. This book offers readers a comprehensive understanding of how stress affects our well-being by examining different aspects of stress and its impact on health. This book is guaranteed to enlighten and inform individuals who are students, researchers, healthcare professionals, or simply individuals with an interest in expanding their knowledge on stress and health. Obtain a copy, explore its contents, and begin a voyage to reveal the captivating realm of stress and its consequences for well-being.

CHAPTER 1

UNDERSTANDING STRESS

Contents

1.1. DEFINITION OF STRESS

A complex phenomenon, stress includes a broad spectrum of emotional, psychological, and physical reactions to both internal and external stimuli. It is generally regarded to be the body's response to any alteration that necessitates a modification or response.

Numerous variables, referred to as stressors, can cause these behaviors. Stressors can be acute or chronic, positive or negative (Yaribeygi et al., 2017). Stress is a natural component of life and is essential to survival since it makes people able to react to dangers or obstacles. However, stress can negatively impact a person's health and well-being if it persists or becomes overpowering (Yaribeygi et al., 2017) (Figure 1.1).

Figure 1.1. *Understanding stress.*

Source: I Got Mind Inc. (2023).

From a physiological standpoint, stress involves the activation of the hypothalamic-pituitary-adrenal (HPA) axis, leading to the release of stress hormones such as cortisol and adrenaline (Turner et al., 2020). These hormones prepare the body for a 'fight-or-flight' response, enhancing physical abilities and mental alertness. Psychologically, stress can manifest as feelings of anxiety, tension, or irritability. As Turner et al. (2020) report, short-term stress can be beneficial by enhancing focus and performance, whereas prolonged exposure to stress can lead to significant health issues, including cardiovascular diseases, mental health disorders, and weakened immune function.

1.2. TYPES OF STRESS

Stress can be categorized into several types based on its nature and duration. When classified based on nature, two main types of stress exist: physiological and psychological stress. According to McCraty & Atkinson (2012), psychological stress is an emotional response to external or internal stimuli that can affect mood, thoughts, and subsequent behavior. Physiological stress, on the other hand, refers to any disturbance to a cell's or organ's homeostasis, leading to a biological response (Campbell & Ehlert, 2012). Some examples of physiological stress include metabolic, heat, and oxidative stress.

When classified based on duration (the commonest classification), the primary types of stress are acute stress, episodic acute stress, and chronic stress (Tran et al., 2020), and are discussed in subsections (Figure 1.2).

Figure 1.2. Three types of stress.

Source: Fletcher (2021).

1.2.1. Acute Stress

The most prevalent type of stress is known as acute stress, which results from particular events or circumstances that are viewed as threatening or challenging. Acute stress is transient and usually goes away once the stressor is removed (Tran et al., 2020). Some examples of acute stress include meeting a deadline, handling a minor accident, and giving a presentation. Although acute stress can cause temporary discomfort, it is generally not harmful if managed effectively.

1.2.2. Episodic Acute Stress

When someone has acute stress episodes on a regular basis, it is called episodic acute stress. People with demanding lifestyles or pessimistic mindsets are more likely to experience this kind of stress. Individuals experiencing episodic acute stress frequently have a lasting sense of crisis, which results in ongoing tension and irritability (Balters et al., 2020). Episodic acute stress might exacerbate chronic health issues like hypertension and cardiovascular disorders if left untreated (Turner et al., 2020).

1.2.3. Chronic Stress

Chronic stress is a type of stress that lasts a long time and is brought on by recurring events or circumstances, such as a tough marriage, protracted financial issues, or a demanding job. Chronic stress can last for weeks, months, or even years, in contrast to acute stress, which is temporary (Turner et al., 2020). Chronic activation of the stress response system can result in a number of health problems, such as immune system weakness, depression, anxiety disorders, and cardiovascular illnesses. Because chronic stress can gradually and silently deteriorate a person's health over time, it is very dangerous.

1.3. CAUSES OF STRESS

Stressors, or the things that lead to stress, can differ greatly from person to person and fall into a number of different categories, such as physiological, psychological, social, and environmental aspects.

1.3.1. Environmental Stressors

As the name suggests, environmental stressors are present in our surroundings and encompass factors such as noise, pollution, natural disasters, and living conditions (McCraty & Atkinson, 2012). For instance, internally displaced persons, refugees, and those living in areas or regions prone to natural disasters may experience heightened levels of stress that surpass the HPA threshold, leading to the health effects of stress. Environmental stressors can be unpredictable and uncontrollable, making them particularly challenging to manage.

1.3.2. Psychological Stressors

Unrealistic expectations, negative self-perception, and internal conflicts are examples of psychological stressors. Stress levels can rise significantly for those who battle perfectionism, low self-esteem, or failure-related anxiety (Turner et al., 2020). Significant life transitions like divorce, a death in the family, or moving can also cause psychological stress.

1.3.3. Social Stressors

Social stressors arise from interpersonal relationships and social dynamics. These include family conflicts, workplace stress, financial difficulties, and social isolation. Relationship issues, such as marital problems or conflicts with colleagues, are common sources of social stress. Additionally, individuals who lack social support or experience social discrimination may be particularly vulnerable to stress (Campbell & Ehlert, 2012).

1.3.4. Physiological Stressors

Physiological stressors are internal factors related to the body's physical condition that can trigger stress responses. These stressors are often related to chronic health conditions, acute illnesses, injuries, and lifestyle factors (Turner et al., 2020). Health problems and physical discomfort can exacerbate stress levels, creating a feedback loop where stress further deteriorates physical health.

1.4. EFFECTS OF STRESS ON THE BODY

Stress has a wide range of impacts on the body, affecting almost every system and resulting in a variety of health problems. The autonomic nerve system and the HPA axis are the main mechanisms that mediate the physiological reactions to stress (Turner et al., 2020) (Figure 1.3).

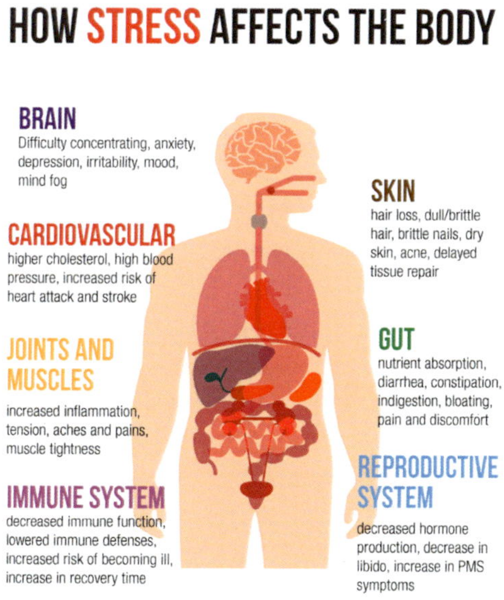

Figure 1.3. *How stress affects the body.*

Source: Body Signals: Stress | Greensboro Chiropractic (2020).

1.4.1. Cardiovascular System

Stress raises blood pressure and heart rate, which has a major negative effect on the cardiovascular system. These things happen once the HPA is activated, which causes the stress hormones—cortisol and adrenaline, in particular—to be secreted. According to Turner et al. (2020), adrenaline is a sympathomimetic, which means that it raises blood pressure and causes consequences like an accelerated heart rate. In contrast, cortisol stimulates the production of low-density lipoproteins (LDLs), which are bad cholesterol that narrow arteries and cause hypertension, atherosclerosis, and a higher risk of heart attacks and strokes (Turner et al., 2020). Additionally, stress-related activities like smoking, eating poorly, and not exercising increase the chances of cardiovascular disease.

1.4.2. Immune System

Stress suppresses the immune system and causes persistent inflammation, which weakens the immune system considerably. Long-term stress increases cortisol levels, which decrease lymphocyte function and production and weaken the body's defenses against infections (Yaribeygi et al., 2017). This is because cortisol activates the hypothalamic-pituitary-adrenal axis. Stress also interferes with the regulation of cytokines, which results in a low-grade inflammation that persists and destroys tissues and organs. According to Yaribeygi et al. (2017), this imbalance worsens autoimmune diseases, prolongs the healing of wounds, and makes people more susceptible to infections. Chronic stress can also impair immunological surveillance, which accelerates the development of illnesses like cancer. As a result, stress seriously impairs immunological response and general health.

1.4.3. The Musculoskeletal System

Stress has a major impact on the musculoskeletal system, mostly by causing discomfort and tension in the muscles. Stress hormones like cortisol and adrenaline are released when a person experiences chronic stress. These chemicals increase muscle tension and prime the body for a "fight-or-flight" reaction (Barsotti et al., 2020). Long-term muscle tension can cause pain and discomfort, especially in the shoulders, back, and neck. This can result in disorders, including migraines and tension headaches.

Furthermore, stress can worsen pre-existing musculoskeletal conditions by lowering the pain threshold and raising sensitivity to pain, such as fibromyalgia and temporomandibular joint (TMJ) dysfunction. According to Barsotti et al. (2020), the repeated activation of these stress reactions might result in musculoskeletal dysfunction, muscle fatigue, and spasms. Additionally, Stress-related habits like bad posture and inactivity can exacerbate these impacts, creating a vicious cycle of discomfort and limited mobility that drastically lowers a person's quality of life.

1.4.4. Endocrine System

Stress causes the release of cortisol and adrenaline, which have a direct impact on the endocrine system, which controls hormones. Long-term high levels of these hormone exposure can interfere with glucose and insulin metabolism, which can lead to diseases including diabetes and obesity (Barsotti et al., 2020). For example, adrenaline surges can impact thyroid function and result in hyper- or hypothyroidism. Chronic stress can also affect reproductive hormones, causing males to have lower testosterone levels and women to have irregular menstrual periods. These hormonal imbalances contribute to various health issues, including obesity, osteoporosis, and reproductive disorders, highlighting the significant impact of stress on endocrine health (Barsotti et al., 2020).

1.4.5. Digestive System

Stress has a profound effect on the digestive system through a number of mechanisms. Firstly, it can trigger the autonomic nervous system, which changes gut motility and secretion and frequently results in gastrointestinal problems like nausea, diarrhea, or constipation (Tran et al., 2020). Secondly, chronic stress can aggravate conditions like irritable bowel syndrome (IBS) and inflammatory bowel disease (IBD) by altering the gut's inflammatory response due to stress hormones like cortisol. Lastly, stress can alter the gut microbiota, which upsets the balance of beneficial bacteria, which is essential for digestion and general gut health. Lastly, stress can affect appetite, potentially leading to overeating or loss of appetite, which can result in fluctuations in weight and nutritional deficiencies (Tran et al., 2020).

1.5. EFFECTS OF STRESS ON THE MIND

As noted by Campbell & Ehlert (2012), stress has a dramatic effect on mental health, affecting behavior, emotional stability, and cognitive capabilities. The psychological repercussions of stress can be equally as crippling as their physical ones.

1.5.1. Cognitive Function

Stress has a severe effect on cognitive performance by weakening critical brain regions and disturbing neural processes. Chronic stress is characterized by elevated cortisol levels, which can harm the hippocampal region and cause memory loss and difficulties creating new memories (Loi et al., 2017). The prefrontal cortex, which controls executive skills including decision-making, attention, and problem-solving, is impacted by stress as well, which lowers cognitive function. Furthermore, neurotransmitter imbalances brought on by stress, especially those involving dopamine and serotonin, are linked to a reduction in focus and an increase in the likelihood of developing mental exhaustion (Loi et al., 2017). Long-term stress exposure can worsen neurodegenerative diseases like Alzheimer's disease and hasten cognitive loss. These consequences underscore the negative impact of stress on cognitive health and the necessity of developing efficient stress management techniques.

1.5.2. Emotional Well-Being

Stress has a significant impact on emotional health and can have a variety of detrimental psychological effects. By dysregulating the hypothalamic-pituitary-adrenal (HPA) axis and changing neurotransmitter levels, particularly serotonin and dopamine, chronic stress can exacerbate anxiety disorders and depression (Loi et al., 2017). This imbalance affects mood, resulting in enduring depressive, irritable, and hopeless sentiments.

Additionally, stress weakens emotional resilience, increasing a person's vulnerability to emotional weariness and burnout. Stress can also affect one's ability to regulate emotions, which can lead to heightened emotional reactivity and trouble controlling annoyance or rage. The detrimental effects of these emotional disturbances on interpersonal relationships and general quality of life highlight how crucial stress management and treatment are to preserving emotional health and well-being.

1.5.3. Behavioral Changes

Stress has a big impact on behavior; it frequently shows up as maladaptive coping mechanisms and routine adjustments. As people look for short-term respite from chronic stress, they may become more dependent on drugs, alcohol, and nicotine, which increases their risk of addiction and associated health issues (Yaribeygi et al., 2017). Stress also throws off sleep cycles, causing hypersomnia or insomnia, which aggravates emotional and mental problems even more. Stress can also change how people eat, leading to either overeating or undereating, which can cause swings in weight and nutritional imbalances. Similar effects are seen in social behaviors; stressed people may become withdrawn, irritable, or aggressive, straining relationships and weakening social support systems (Yaribeygi et al., 2017). These modifications in behavior highlight the widespread effects of stress on general health and day-to-day functioning.

1.6. COPING MECHANISMS FOR STRESS

In order to control stress and lessen its negative effects on health, effective coping strategies are crucial. Problem-focused, emotion-focused, and adaptive coping strategies can be used to group the various coping techniques.

1.6.1. Problem-Focused Coping

As Tran et al. (2020) report, problem-focused coping aims to decrease or eliminate stress by directly addressing its source. This strategy works well in situations where people have some influence over the stressor. Among the strategies are:

1. **Time Management:** Prioritizing tasks and setting realistic goals to reduce workload and pressure.
2. **Problem-Solving:** Identifying the problem, generating possible solutions, and implementing the most effective one.
3. **Seeking Support:** Asking for help from friends, family, or professionals to manage stressors.

1.6.2. Emotion-Focused Coping

Instead of focusing on the stressor itself, Tran et al. (2020) describe emotion-focused coping which attempts to control the emotional reaction to stress. When one cannot control the stressor, this strategy is helpful. Some methods of emotion-focused coping include:

1. **Relaxation Techniques:** Practices such as deep breathing, meditation, and progressive muscle relaxation to reduce physiological arousal.
2. **Cognitive Reframing:** Changing negative thought patterns and perceptions to reduce the emotional impact of stress.
3. **Mindfulness:** Focusing on the present moment and accepting it without judgment to reduce anxiety and enhance emotional regulation.

1.6.3. Adaptive Coping Mechanism

Described by Tran et al. (2020) as healthy strategies that promote resilience and well-being, adaptive stress coping mechanisms develop into habits that serve to distract us from life stressors. They include:

1. **Physical Activity:** Regular exercise helps reduce stress hormones, improve mood, and boost overall health.
2. **Healthy Diet:** Eating a balanced diet supports physical and mental health, providing the body with essential nutrients.

3. **Adequate Sleep:** Ensuring sufficient and quality sleep is crucial for recovery and stress management.
4. **Social Connections:** Maintaining strong social relationships provides emotional support and a sense of belonging.

1.6.4. Professional Intervention

Effective stress management may occasionally require expert assistance. Counseling, mindfulness-based stress reduction (MBSR), cognitive-behavioral therapy (CBT), and other therapies can give people coping mechanisms and strategies. Severe disorders connected to stress may also require prescription medication (McCraty & Atkinson, 2012).

CHAPTER 2

THE LINK BETWEEN STRESS AND DISEASE

Contents

2.1. INTRODUCTION

Stress results from a variety of social, psychological, and environmental factors and is an unavoidable part of living. Acute stress is an essential survival strategy for the body, but persistent stress can be harmful to health. As we discussed in the last chapter, a protracted state of psychological tension characterizes chronic stress, and research indicates that it plays a major role in the onset and course of many diseases. Recent studies have demonstrated the significant negative effects of long-term stress on mental and physical health (Turner et al., 2020; Loi et al., 2017).

The hypothalamic-pituitary-adrenal (HPA) axis and the sympathetic nervous system are activated in the physiological processes that underlie the stress-disease relationship. This results in the prolonged release of stress hormones, including cortisol and adrenaline. Though beneficial during brief periods of stress, these hormonal shifts turn maladaptive when they persist, causing immune dysregulation, metabolic disruptions, and systemic inflammation (Barsotti et al., 2020). A large number of illnesses, including gastrointestinal disorders, anxiety, and depression, cardiovascular diseases, and mental health problems, are largely caused by this dysregulation.

Moreover, behavioral factors modulate the link between disease and stress. According to Cohen et al. (2016), persistent stress frequently results in unhealthy coping strategies like drinking alcohol, smoking, and adopting bad eating habits, all of which increase the chance of developing chronic illnesses. The intricate relationship between stress's effects on behavior, psychology, and biology highlights how complicated these relationships are.

Numerous research demonstrating the complex nature of the interaction between stress and disease have been inspired by significant breakthroughs in our understanding of the stress-disease nexus (Barsotti et al., 2020; Turner et al., 2020; Yaribeygi et al., 2017). These results highlight the role that stress management therapies play in both preventing and reducing disease, opening the door for more integrated approaches to healthcare that take mental and physical health into account.

2.2. CHRONIC STRESS AND DISEASE

Unlike acute stress, chronic stress is characterized by prolonged exposure to stressors without adequate downtime or recuperation. The hypothalamic-pituitary-adrenal (HPA) axis and the autonomic nervous system are stimulated by this ongoing stress, which results in the constant release of stress hormones, including cortisol and adrenaline (Barsotti et al., 2020). Although these hormones are necessary for the "fight or flight" reaction, prolonged elevation of their levels can be harmful and play a significant role in the pathogenesis and prognosis of chronic diseases such as diabetes, high blood pressure (hypertension), and obesity, among others.

As Yaribeygi et al. (2017) mention, chronic stress affects almost all body tissues and organs leading to multiple health conditions as highlighted below.

2.2.1. Impacts on the Immune System

Chronic stress refers to long-term, continuous stressors that significantly impair the immune system's capacity to function. The immune system is extremely sensitive to psychological stress and is an essential component of the body's defense against infections and illnesses. Numerous studies have examined the connection between long-term stress and immune function (Cohen et al., 2016; Barsotti et al., 2020; Yaribeygi et al., 2017), illuminating the intricate processes by which stress hormones affect immunological reactions.

The HPA axis is a major channel via which the immune system is impacted by chronic stress. Cortisol and other glucocorticoids are raised when the HPA axis is activated for an extended period (Barsotti et al., 2020). Although cortisol is necessary for controlling immunological responses, long-term increases in it can have immunosuppressive effects. High cortisol levels decrease the efficiency of the immune response because they suppress cytokine generation and activity, which are proteins essential for immune system communication. This reduction in cytokine production can make the body more vulnerable to disease and hinder its capacity to develop a successful defense against diseases.

The distribution and function of immune cells are also impacted by long-term stress. Studies reveal that long-term stress can cause a decrease in the number and capacity of lymphocytes, such as T and B cells, which are essential for adaptive immunity (Barsotti et al., 2020). Furthermore, changes in leukocyte trafficking brought on by stress may lead to a reduction in these cells' concentration in vital body regions, thus impairing immune function.

Moreover, prolonged stress also significantly impacts the immune system, as it can lead to systemic inflammation. Chronic stress triggers the release of catecholamines like noradrenaline and adrenaline by activating the sympathetic-adrenal-medullary (SAM) axis. These catecholamines contribute to a chronic low-grade inflammatory state by increasing the synthesis of pro-inflammatory cytokines (McEwen, 2017). Many diseases, such as diabetes, heart disease, and some types of cancer, have been linked to chronic inflammation in their early stages.

Furthermore, chronic stress has been associated with the dysregulation of natural killer (NK) cells, which are essential for the body's defense against tumors and virally infected cells. Studies have shown that chronic stress can decrease NK cell activity, thereby impairing the body's ability to identify and eliminate malignant cells (Dhabhar, 2014; McEwen, 2017). This reduction in NK cell function can increase the risk of cancer development and progression.

Chronic stress can also cause behavioral changes that impair immune function in addition to these physiological alterations. According to Turner et al. (2020), stress frequently leads to poor sleep, improper eating habits, and decreased physical exercise, all of which can exacerbate immune system weakness.

2.3. CARDIOVASCULAR HEALTH AND CHRONIC STRESS

Given their high prevalence and substantial impact on mortality and morbidity, cardiovascular diseases are a major public health concern. The mechanisms through which stress influences cardiovascular health are complex and multifaceted, involving behavioral, physiological, and biochemical pathways (Figure 2.1).

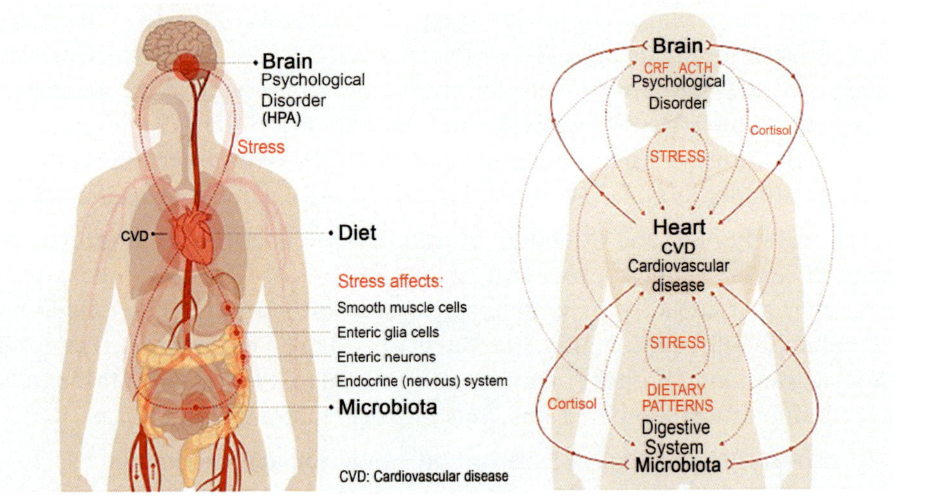

Figure 2.1. *Relationship between nutritional changes and emotional state and brain health.*

Source: Morera et al. (2019).

The cardiovascular system is particularly vulnerable to the effects of chronic stress and may develop the conditions described below.

2.3.1. Hypertension

High blood pressure, often known as hypertension, is a prevalent cardiovascular disease that has a strong correlation to ongoing stress. Numerous processes involving the neurological and endocrine systems cause stress to have an impact on blood pressure. The body releases stress hormones like cortisol and catecholamines (epinephrine and norepinephrine) when it experiences stress via stimulating the hypothalamic-pituitary-adrenal (HPA) and sympathetic-adrenal-medullary (SAM) axes. These hormones speed up the heartbeat and constrict blood vessels, which momentarily elevates blood pressure (Kivimäki & Steptoe, 2018).

On the other hand, prolonged exposure to these hormones as a result of chronic stress might result in persistently elevated blood pressure. Chronic activation of the SAM and HPA axes leads to higher heart rate and persistent vascular resistance, which exacerbates hypertension (McEwen, 2017; Kivimäki & Steptoe, 2018). Furthermore, stress can lead to poor food choices, inactivity, and increased alcohol and tobacco use—all of which are risk factors for hypertension (Kivimäki & Steptoe, 2018). Furthermore, changes in the renin-angiotensin system, which is essential for controlling blood pressure, have been connected to long-term stress. According to McEwen (2017), dysregulation of this system might result in fluid retention and elevated vascular resistance, which exacerbate hypertension. Apart from heart-related issues, high blood pressure poses a significant risk for cerebrovascular illnesses. Prolonged elevated blood pressure can harm brain blood vessels, raising the risk of stroke and transient ischemic episodes. Severely increased blood pressure can also result in hypertensive encephalopathy, which is marked by headaches, confusion, and even seizures (Kivimäki & Steptoe, 2018).

Hypertension also has negative effects on the renal system. Through fluid and electrolyte balance, the kidneys control blood pressure; nevertheless, persistently high blood pressure can harm the renal arteries, decreasing blood supply to the kidneys and compromising their ability to function. Renal failure may result from this illness if it continues to worsen and develop into chronic kidney disease (McEwen, 2017).

Moreover, hypertension can cause hypertensive retinopathy, which damages the eyes. This disorder affects the blood vessels in the retina, which may lead to loss or impairment of vision. Peripheral artery disease risk is increased by atherosclerosis, a disorder marked by the hardening and constriction of the arteries, which is also a result of long-term uncontrolled hypertension (McEwen, 2017). Because hypertension has a wide range of complex and systemic impacts, it is crucial to identify the condition early, treat it effectively, and change one's lifestyle to reduce risks and enhance long-term health outcomes.

2.3.2. Atherosclerosis

An increasing number of studies have identified chronic stress as a major risk factor for the onset and course of atherosclerosis, a disorder marked by the build-up of plaques rich in fat within artery walls. Cardiovascular conditions like peripheral artery disease, coronary artery disease, and stroke can be brought on by this pathological process. There are several different ways that chronic stress affects atherosclerosis, including immunological, neuroendocrine, and behavioral routes.

Chronic stress stimulates the hypothalamic-pituitary-adrenal (HPA) axis, which causes a prolonged release of glucocorticoids, especially cortisol, at the neuroendocrine level. Extended exposure to high cortisol levels can have a number of detrimental effects on the cardiovascular system. For example, cortisol can increase endothelial dysfunction, which is a risk factor for atherosclerosis, by decreasing nitric oxide's bioavailability, a vital vasodilator. According to Kivimäki & Steptoe (2018), endothelial dysfunction promotes inflammation and the subsequent development of atherosclerotic plaques by facilitating leukocyte adherence to the vascular endothelium.

Additionally, long-term stress triggers the sympathetic nervous system (SNS), which raises catecholamine levels like noradrenaline and adrenaline. These hormones cause hypertension and hypercoagulability, two conditions that increase the risk of atherosclerosis. Furthermore, catecholamines have the ability to increase macrophages' absorption of low-density lipoprotein (LDL) cholesterol in artery walls, causing them to become foam cells and facilitating the development of plaque (Kivimäki & Steptoe, 2018).

In circumstances of prolonged stress, the immune system is essential to the pathophysiology of atherosclerosis. Tumor necrosis factor-alpha (TNF-α) and interleukin-6 (IL-6) are two pro-inflammatory cytokines that are overexpressed during times of stress. These cytokines encourage immune cell recruitment and activation, including T-lymphocytes and monocytes, in the artery intima, where they add to the inflammatory environment that fuels atherogenesis. Persistent stress can also hinder the process of inflammation resolution, which can result in the persistence and advancement of atherosclerotic lesions (Liu et al., 2017).

Chronic stress-related behavioral factors raise the risk of atherosclerosis even more. Maladaptive habits including smoking, excessive alcohol consumption, poor diet, and physical inactivity can all be brought on by stress and are known risk factors for atherosclerosis. Furthermore, sleep disturbances brought on by stress can exacerbate metabolic dysregulation, such as dyslipidemia and insulin resistance, which accelerate the development of atherosclerotic plaques (Kivimäki & Steptoe, 2018). In summary, a complicated interaction between neuroendocrine activation, immune system modulation, and behavioral alterations leads to chronic stress's role in atherosclerosis. The significance of lifestyle modifications and stress management in the prevention and treatment of atherosclerotic cardiovascular diseases is highlighted by an understanding of these mechanisms.

2.3.3. Coronary Artery Disease

Atherosclerosis-related narrowing or blockage of the coronary arteries is the hallmark of coronary artery disease (CAD). There are several different ways that chronic stress and CAD are related, including direct physiological consequences of stress and behavioral changes brought on by stress. According to Kivimäki & Steptoe (2018), long-term stress causes the HPA and SAM axis to remain activated, which raises heart rate, blood pressure, and vascular inflammation and all of these factors contribute to the development of coronary artery disease (CAD).

Chronic stress causes the HPA axis to remain activated, which raises levels of glucocorticoids, especially cortisol, in the neuroendocrine system. Chronic rise of cortisol is linked to metabolic disorders, including insulin resistance, dyslipidemia, and hyperglycemia, all of which are known risk factors for coronary heart disease (CAD). Additionally, cortisol increases the pro-inflammatory state inside the vascular endothelium and reduces the bioavailability of nitric oxide, which impairs vasodilation and causes endothelial dysfunction (McEwen, 2017).

Chronic stress also increases the release of catecholamines, such as noradrenaline and adrenaline, by activating the sympathetic nervous system (SNS). The danger of myocardial ischemia is increased by this prolonged sympathetic activity, which also causes tachycardia, hypertension, and an increased oxygen demand in the heart. Moreover, catecholamines raise blood viscosity and platelet aggregation, which causes a hypercoagulable condition and raises the risk of thrombus formation and subsequent coronary events (Turner et al., 2020).

The pathophysiology of CAD is largely dependent on inflammation, and prolonged stress greatly exacerbates inflammatory processes. The production of pro-inflammatory cytokines, such as TNF-α and interleukin-6 (IL-6), is triggered by stress and facilitates the recruitment of T-lymphocytes and monocytes to the endothelium of the coronary artery. Atherosclerotic plaque development and progression are facilitated by the infiltration of these immune cells into the arterial wall. According to Dhabhar (2014), a persistent inflammatory milieu increases the risk of acute coronary events and plaque rupture by exacerbating plaque instability. The risk of CAD is significantly increased by behavioral reactions to ongoing stress. Maladaptive behaviors including smoking, excessive alcohol consumption, poor eating habits, and physical inactivity are frequently brought on by stress. Obesity, hypertension, and dyslipidemia are all major risk factors for CAD that are facilitated by these habits. Moreover, sleep disorders brought on by stress may intensify these behavioral and metabolic risk factors, resulting in a vicious cycle that carries on cardiovascular risk (Liu et al., 2017).

2.4. IMMUNE SYSTEM FUNCTION AND CHRONIC STRESS

Stress has a profound influence on the immune system in a number of ways, from increased inflammation to immunosuppression. A complicated interaction between immunological and neuroendocrine responses mediates the link between stress and immune function (Figure 2.2). Stress affects immune function in several ways which are discussed in subsections.

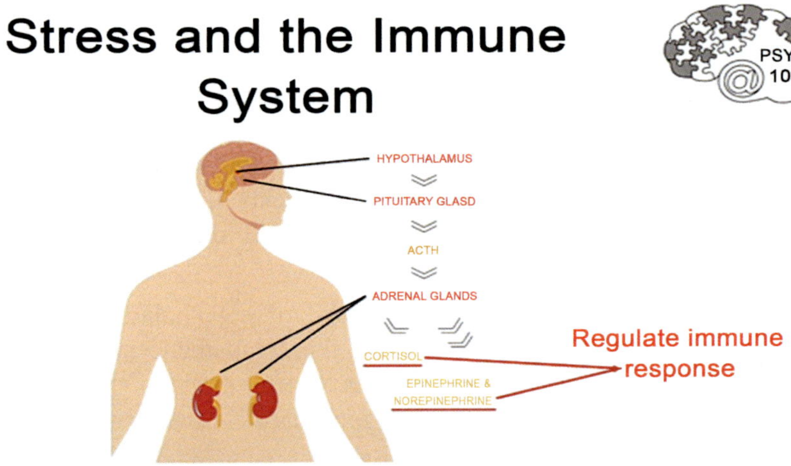

Figure 2.2. *Stress and the immune system.*

2.4.1. Immunosuppression

Stress has a well-documented effect on the body's immunological response, which has important consequences for both health and illness. As previously mentioned, chronic stress causes a continuous release of stress hormones including cortisol and catecholamines due to prolonged stimulation of the sympathetic nervous system and hypothalamic-pituitary-adrenal (HPA) axis. Although these hormones are essential for immediate stress reactions, prolonged elevation of their levels can compromise immunological function (Turner et al., 2020).

The main glucocorticoid secreted during stress, cortisol, has a strong immune-suppressive effect. By attaching to glucocorticoid receptors on immune cells, it affects cellular activity and gene expression, which in turn controls immunological function. Suppression of lymphocyte proliferation is one of cortisol's most important effects. T cells and B cells are examples of lymphocytes, which are crucial for adaptive immune responses. By causing apoptosis, which lowers their quantity and hinders the body's capacity to mount a potent immunological response, cortisol prevents their proliferation (Liu et al., 2017).

Moreover, cortisol influences the synthesis of cytokines, which are essential for immune cell coordination and communication. It inhibits the synthesis of pro-inflammatory cytokines like TNF-α, interleukin-6 (IL-6), and interleukin-1 (IL-1) and increases the release of anti-inflammatory cytokines like interleukin-10 (IL-10). The immune system's capacity to respond to infections and tissue damage is suppressed by this change towards an anti-inflammatory profile, making people more susceptible to sickness (Liu et al., 2017). Natural killer (NK) cells, an essential part of the innate immune system in charge of identifying and removing malignant and virally infected cells, are also impacted by long-term stress. The body's first line of defense against infections and cancers can be weakened by stress hormones, especially norepinephrine, which can lower NK cell activity and cytotoxic capability (Turner et al., 2020; Liu et al., 2017).

Persistent stress can affect immune cell trafficking and distribution in addition to its direct hormonal effects. Immune cells may be redistributed from peripheral tissues back into the bloodstream as a result of stress-induced alterations in blood flow and adhesion molecule expression. This weakens immune protection by decreasing the quantity of immune cells at infection or damage sites (Dhabhar, 2014).

Long-term effects on immunological memory are also a result of persistent HPA axis activation and the glucocorticoid increase that goes along with it. According to studies, long-term stress can hinder the development and upkeep of immune memory, reducing the efficacy of immunizations and raising the chance of recurrent infections (Cohen et al., 2016).

Chronic stress also affects other physiological systems, which in turn affects how the immune system is regulated. For example, stress can interfere with sleep cycles, and it is well-recognized that getting too little sleep lowers immunity. According to Liu et al. (2017), stress can also result in harmful habits including substance misuse, poor eating, and decreased physical activity, all of which have a detrimental effect on the immune system.

2.4.2. Increased Inflammation

Changes in cytokine production are a major way that stress exacerbates inflammation. Signaling proteins called cytokines control inflammatory and immunological reactions. The body's initial reaction to acute stress usually entails a brief rise in pro-inflammatory cytokines, which aid in priming the body to handle any damage or infection. Chronic stress, however, upsets this equilibrium and causes a persistent rise in pro-inflammatory cytokine production, including interleukin-6 (IL-6), tumor necrosis factor-alpha (TNF-α), and interleukin-1 beta (IL-1β) (Cohen et al., 2016) (Figure 2.3).

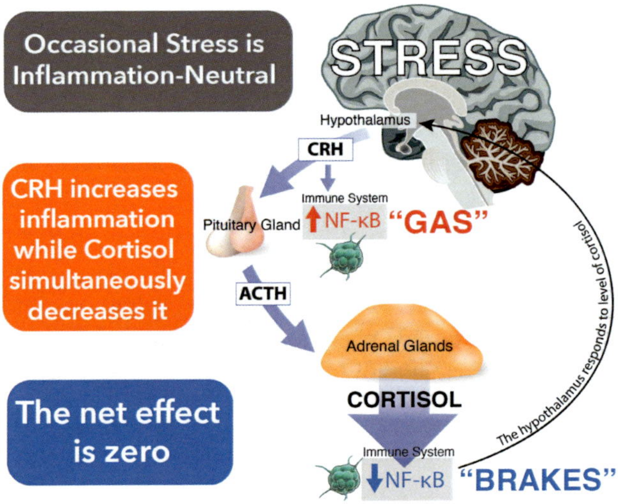

Figure 2.3. *How does stress affect your immunity?*

Source: Dorsey (2022).

The sustained rise of glucocorticoids such as cortisol is responsible for the dysregulation of cytokine production under chronic stress. Cortisol normally has an anti-inflammatory impact, but long-term exposure causes immune cells to develop glucocorticoid resistance, which reduces their reactivity to cortisol's regulating effects. As a result, pro-inflammatory cytokines are produced uncontrollably (Miller et al., 2002). Furthermore, the synthesis of these cytokines can be further enhanced by catecholamines generated during stress, such as norepinephrine, which can intensify the inflammatory response (Liu et al., 2017).

The activation of nuclear factor kappa B (NF-κB), a crucial transcription factor in the inflammatory response, is another way that stress-induced inflammation is mediated. Genes implicated in inflammation are expressed more when NF-κB is activated by prolonged stress. According to Liu et al. (2017), this activation contributes to a chronic inflammatory state by continuing the production of pro-inflammatory cytokines and other inflammatory mediators.

Furthermore, immune cell activity is impacted by stress. Chronic stress causes immune cells such as monocytes and macrophages to become more mobilized and activated. This activation leads to the release of pro-inflammatory cytokines and reactive oxygen species (ROS), which further exacerbates inflammation (Liu et al., 2017). Many inflammatory and autoimmune diseases, including rheumatoid arthritis, cardiovascular disorders, and depression, can arise and worsen as a result of this increased inflammatory response.

2.4.3. Delayed Wound Healing

Stress primarily impacts wound healing by dysregulating glucocorticoids, especially cortisol. Cortisol normally aids in regulating immunological responses and inflammation. On the other hand, long-term stress raises cortisol levels, which can inhibit important steps in the cascade of wound healing. Increased cortisol prevents keratinocytes and fibroblasts from proliferating, which are necessary for collagen synthesis and re-epithelialization, respectively. New tissue at the wound site forms more slowly as a result of this inhibition (Liu et al., 2017).

Chronic stress also modifies cytokine production and function, which are important proteins for immune response coordination during wound healing. An imbalance between pro- and anti-inflammatory cytokines might result from stress. For example, persistent inflammation might impede the healing process due to the overproduction of pro-inflammatory cytokines including interleukin-6 (IL-6) and tumor necrosis factor-alpha (TNF-α). Conversely, increased cortisol levels have the ability to inhibit the synthesis of cytokines, such as interleukin-1 (IL-1) and interferon-gamma (IFN-γ), which are essential for initiating the inflammatory response that gets the wound ready to heal (Dhabhar, 2014).

Angiogenesis, the process of creating new blood vessels that are essential for delivering nutrients and oxygen to the healing tissue, is similarly impacted by stress. By decreasing the expression of vascular endothelial growth factor (VEGF), a crucial modulator of the development of new blood vessels, chronic stress hinders angiogenesis (Dhabhar, 2014). Insufficient blood flow causes the repairing tissue to get insufficient oxygen and nutrients, which further slows down the healing process.

Stress also affects the immune cells' recruitment and functionality at the site of the wound. For example, under prolonged stress circumstances, macrophages,which are essential for wound healing—are less effective in removing debris and pathogens and secreting growth hormones. Slower tissue regeneration and less effective wound clearance are the outcomes of this reduced macrophage function (Liu et al., 2017). Moreover, through behavioral alterations, psychological stress also has an indirect impact on the healing of wounds. Stress can cause bad eating habits, insufficient exercise, and restless nights, all of which can impair immune system performance and tissue healing mechanisms (Yaribeygi et al., 2017).

2.5. DIGESTIVE SYSTEM ISSUES AND STRESS

Stress and digestive system problems have a complex association that has attracted a lot of research in the medical and psychological domains. This section examines the physiological mechanisms underlying the effects of stress on the digestive system, as well as typical digestive problems associated with stress and management techniques for these conditions (Figure 2.4).

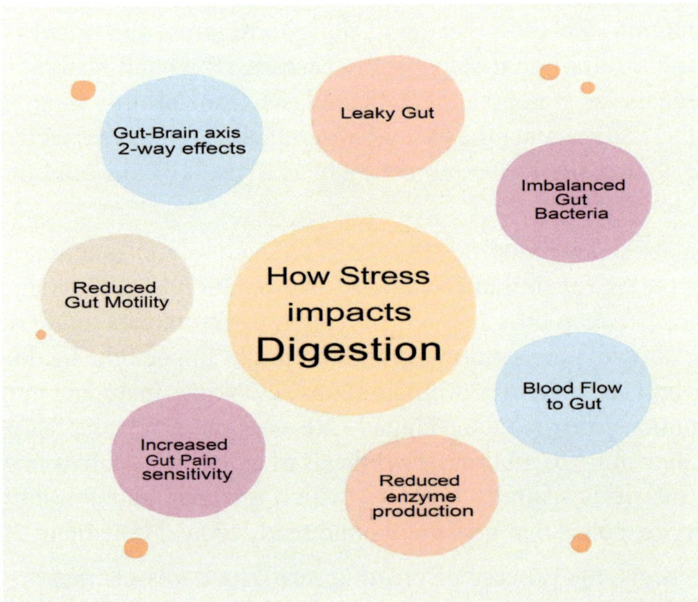

Figure 2.4. *Stress and digestion.*

Source: Brunt (2023).

2.5.1. The Physiology of Stress and the Digestive System

The "fight or flight" response is a series of physiological reactions that stress is a normal reaction to when it perceives threats or difficulties. The sympathetic nervous system, which releases stress hormones including cortisol and adrenaline, is the main player in this reaction (Schneiderman et al., 2005). Although this process is necessary for survival, prolonged stress can cause chronic activation, which can be harmful to the digestive system and other body systems. The digestive system, comprising the gastrointestinal (GI) tract and accessory organs like the liver, pancreas, and gallbladder, is highly sensitive to stress. The gut is often referred to as the "second brain" because of the enteric nervous system, a vast network of neurons that governs its function independently yet communicates with the central nervous system (Schneiderman et al., 2005). This gut-brain axis plays a crucial role in maintaining digestive health and is significantly influenced by stress.

2.5.2. Stress-Induced Digestive Disorders

Several digestive disorders are closely linked to stress, with irritable bowel syndrome (IBS) being one of the most prevalent. IBS is a functional GI disorder characterized by symptoms like abdominal pain, bloating, and altered bowel habits without any identifiable organic cause. As Yaribeygi et al. (2017) state, stress can exacerbate these symptoms by affecting gut motility, increasing intestinal permeability, and altering the gut microbiota.

Gastroesophageal reflux disease (GERD) is another condition where stress plays a critical role. GERD occurs when stomach acid frequently flows back into the esophagus, causing symptoms such as heartburn and regurgitation (Chu et al., 2024). Furthermore, stress can worsen GERD by increasing acid production and compromising the esophageal sphincter's ability to prevent acid reflux.

Stress is also implicated in the development and exacerbation of peptic ulcers. While the primary cause of peptic ulcers is infection with Helicobacter pylori bacteria, stress can aggravate the condition by increasing gastric acid secretion and reducing the stomach's protective mucus layer. Furthermore, stress can delay the healing of existing ulcers and increase the risk of complications (Yaribeygi et al., 2017).

Further research by Yaribeygi et al. (2017) indicates that inflammatory bowel disease (IBD), which includes Crohn's disease and ulcerative colitis, is another area where stress has a profound impact. Although IBD is primarily driven by immune dysregulation, stress can trigger flare-ups and worsen the severity of symptoms. The mechanisms include altered immune responses, increased intestinal permeability, and changes in gut microbiota composition.

2.5.3. Mechanisms Linking Stress and Digestive Health

Understanding the mechanisms by which stress affects the digestive system is crucial for developing effective interventions. One key mechanism is the alteration of gut motility. Stress can either accelerate or slow down gut transit time, leading to diarrhea or constipation, respectively (Chu et al., 2024). This is mediated by stress hormones like cortisol, which influence the enteric nervous system and smooth muscle function in the GI tract.

Another mechanism involves increased intestinal permeability, often referred to as "leaky gut." Chronic stress can weaken the tight junctions between intestinal epithelial cells, allowing pathogens, toxins, and partially digested food particles to enter the bloodstream ('Impact of Stress, Immunity, and Signals from Endocrine and Nervous System on Fascia,' 2021). This can trigger immune responses and systemic inflammation, contributing to digestive and other systemic diseases.

The gut microbiota, a diverse community of microorganisms residing in the intestines, also plays a pivotal role in the gut-brain axis. Stress can disrupt the balance of these microbial communities, leading to dysbiosis. Dysbiosis has been linked to various digestive disorders, including IBS and IBD. The altered microbiota can influence gut motility, immune responses, and the production of neurotransmitters like serotonin, which modulate mood and GI function (Yaribeygi et al., 2017).

2.5.4. Management Strategies for Stress-Related Digestive Disorders

Managing stress-related digestive disorders requires a multifaceted approach that addresses both psychological and physiological aspects. Stress management techniques such as cognitive-behavioral therapy (CBT), mindfulness-based stress reduction (MBSR), and relaxation exercises have shown efficacy in alleviating symptoms of IBS, GERD, and other stress-related GI conditions (Chu et al., 2024). These therapies help individuals develop coping strategies, reduce perceived stress, and improve overall mental well-being.

Dietary modifications are also crucial in managing stress-related digestive disorders. A balanced diet rich in fiber, probiotics, and anti-inflammatory foods can support gut health and alleviate symptoms. For instance, fiber aids in regulating bowel movements, while probiotics promote a healthy gut microbiota. Additionally, avoiding trigger foods that exacerbate symptoms, such as spicy foods, caffeine, and alcohol, can be beneficial.

Pharmacological treatments may be necessary for some individuals, particularly those with severe symptoms. Medications such as antispasmodics, acid reducers, and anti-inflammatory drugs can provide symptomatic relief and improve quality of life (Schneiderman et al., 2005). However, these should be used under medical supervision, considering potential side effects and interactions.

Regular physical activity is another effective strategy for managing stress and improving digestive health. Exercise stimulates the release of endorphins, which enhance mood and reduce stress. Additionally, physical activity promotes gut motility and reduces the risk of constipation (McCraty & Atkinson, 2012). Practices like yoga and tai chi combine physical movement with mindfulness, offering a holistic approach to stress reduction and digestive health.

Lastly, maintaining a healthy lifestyle that includes adequate sleep, hydration, and social support is essential for managing stress and its impact on the digestive system. Sleep is particularly important, as it allows the body to repair and restore itself, including the digestive system. Hydration helps maintain mucosal linings in the GI tract and facilitates digestion, while social support provides emotional resilience and reduces perceived stress.

The interplay between stress and digestive system issues is complex and multifaceted, involving physiological, psychological, and lifestyle factors. Understanding this relationship is crucial for developing effective management strategies that address both the mind and body. By adopting a holistic approach that includes stress management, dietary modifications, pharmacological treatments, physical activity, and healthy lifestyle practices, individuals can mitigate the impact of stress on their digestive health and improve their overall well-being. As research continues to unravel the intricacies of the gut-brain axis, new insights and interventions will emerge, offering hope for those affected by stress-related digestive disorders.

2.6. MENTAL HEALTH DISORDERS AND STRESS

Mental health disorders and stress are intricately linked, often forming a vicious cycle where each exacerbates the other. Understanding the nature of this relationship, the mechanisms behind it, and the strategies for managing both is essential for improving overall mental well-being. This section will explore the various types of mental health disorders, the role of stress in these conditions, and effective methods for coping and treatment.

2.6.1. Types of Mental Health Disorders

As Kendrick & Pilling (2012) state, mental health disorders encompass a wide range of conditions that affect mood, thinking, and behavior. Common mental health disorders include:

1. **Anxiety Disorders:** It is characterized by excessive fear or anxiety, these disorders include generalized anxiety disorder (GAD), panic disorder, and social anxiety disorder. Symptoms often include restlessness, fatigue, difficulty concentrating, and physical symptoms such as increased heart rate.

2. **Depressive Disorders:** These involve persistent feelings of sadness, loss of interest in activities, and a range of physical and emotional problems. Major depressive disorder (MDD) and dysthymia are prime examples, with symptoms like fatigue, changes in appetite, and suicidal thoughts.
3. **Bipolar Disorder:** This disorder causes extreme mood swings that include emotional highs (mania or hypomania) and lows (depression). During manic phases, individuals may feel euphoric, overly energetic, and impulsive, while depressive phases mirror symptoms of MDD.
4. **Schizophrenia Spectrum and Other Psychotic Disorders:** These disorders are characterized by distorted thinking and awareness, often leading to delusions, hallucinations, and disorganized speech or behavior.
5. **Obsessive-Compulsive Disorder (OCD):** This involves unwanted and intrusive thoughts (obsessions) and repetitive behaviors or mental acts (compulsions) that individuals feel driven to perform.
6. **Post-Traumatic Stress Disorder (PTSD):** This develops after exposure to a traumatic event, leading to symptoms such as flashbacks, nightmares, severe anxiety, and uncontrollable thoughts about the event.

2.6.2. The Role of Stress in Mental Health Disorders

Stress is a natural response to challenging situations and is essential for survival. However, chronic or excessive stress can lead to or exacerbate mental health disorders. Stress affects the brain and body in several ways:

1. **Biological Mechanisms:** Stress triggers the release of hormones such as cortisol and adrenaline. While these hormones help manage short-term stress, chronic stress can lead to an overproduction of cortisol, which is linked to anxiety, depression, and other mental health issues. Long-term stress can alter brain structure and function, particularly in areas related to memory and emotion regulation, like the hippocampus and amygdala (Schneiderman et al., 2005).
2. **Psychological Impact:** Stress can cause or worsen mental health disorders through cognitive and emotional pathways. Persistent stress can lead to negative thought patterns, reduced self-esteem, and a sense of helplessness, all of which contribute to mental health problems. Individuals with pre-existing mental health disorders may experience heightened symptoms during stressful periods.

3. **Behavioral Consequences:** Stress often leads to unhealthy coping mechanisms, such as substance abuse, overeating, or social withdrawal, which can further deteriorate mental health (McCraty & Atkinson, 2012). For example, individuals with anxiety might avoid social situations, leading to increased isolation and worsening symptoms.

2.6.3. Coping with Stress and Mental Health Disorders

Managing stress and mental health disorders requires a multifaceted approach that includes lifestyle changes, therapeutic interventions, and sometimes medication. Here are some effective strategies:

1. Therapeutic Interventions:
   - i. **Cognitive Behavioral Therapy (CBT):** This form of therapy helps individuals identify and change negative thought patterns and behaviors. It is particularly effective for anxiety and depressive disorders.
   - ii. **Dialectical Behavior Therapy (DBT):** A type of CBT that focuses on emotion regulation and interpersonal effectiveness, DBT is often used for borderline personality disorder and other conditions involving emotional dysregulation.
   - iii. **Exposure Therapy:** Often used for PTSD and phobias, this therapy involves gradually exposing individuals to feared situations to reduce anxiety over time.
   - iv. **Mindfulness-based Stress Reduction (MBSR):** This program teaches mindfulness meditation to reduce stress and improve psychological well-being (Tran et al., 2020).
2. Medication:
   - i. **Antidepressants:** Medications like selective serotonin reuptake inhibitors (SSRIs) are commonly prescribed for depression and anxiety.
   - ii. **Anxiolytics:** Medications such as benzodiazepines can help manage severe anxiety, though they are generally used short-term due to the risk of dependency.
   - iii. **Mood Stabilizers:** Used to treat bipolar disorder, these medications help balance mood swings.
   - iv. **Antipsychotics:** These medications are used to treat schizophrenia and other psychotic disorders (Chu et al., 2024).

3. Lifestyle Changes:
   - i. **Regular Exercise:** Physical activity has been shown to reduce symptoms of depression and anxiety. Exercise releases endorphins, which are natural mood lifters, and helps reduce the levels of stress hormones (Schneiderman et al., 2005).
   - ii. **Healthy Diet:** A balanced diet rich in fruits, vegetables, and whole grains supports brain health and overall well-being.
   - iii. **Adequate Sleep:** Poor sleep can exacerbate mental health disorders, so establishing a regular sleep schedule and creating a restful environment are crucial.
   - iv. **Stress Management Techniques:** Practices such as yoga, meditation, and deep-breathing exercises can help reduce stress levels and improve mental health (Tran et al., 2020).
4. **Social Support:** Building strong relationships and seeking support from friends, family, or support groups can provide emotional comfort and practical assistance. Social connections are crucial for mental health, helping individuals feel understood and less isolated.
5. **Professional Support:** Consulting with mental health professionals such as psychologists, psychiatrists, and counselors can provide personalized treatment plans and ongoing support (Tran et al., 2020). Professional guidance is particularly important for severe or complex mental health disorders.

The interplay between mental health disorders and stress underscores the importance of a holistic approach to mental health care. By understanding the types of mental health disorders, recognizing the impact of stress, and implementing effective coping strategies, individuals can improve their mental well-being and lead more fulfilling lives. It is crucial to destigmatize mental health issues and promote awareness so that more people feel empowered to seek help and support. As research and public understanding continue to evolve, the hope is that more effective treatments and preventative measures will emerge, ultimately reducing the burden of mental health disorders and stress on individuals and society.

2.7. STRESS-RELATED DISORDERS

Stress is a ubiquitous aspect of modern life, affecting individuals across all demographics. While a certain level of stress can be beneficial, acting as a motivator, chronic or intense stress can lead to significant health issues. Stress-related disorders encompass a range of mental health conditions triggered or exacerbated by prolonged stress. These disorders include anxiety disorders, depression, post-traumatic stress disorder (PTSD), and psychosomatic conditions. Understanding these disorders, their causes, symptoms, and management strategies is crucial for mitigating their impact on individuals and society (Shalev, 2009).

Stress is the body's response to any demand or threat, real or perceived. This response, known as the "fight or flight" reaction, involves the release of stress hormones such as cortisol and adrenaline, preparing the body for immediate action. While this reaction is beneficial in short bursts, chronic stress can have detrimental effects on physical and mental health.

1. **Anxiety Disorders:** These are among the most common stress-related disorders. They include generalized anxiety disorder (GAD), panic disorder, social anxiety disorder, and specific phobias (Shalev, 2009). These conditions are characterized by excessive fear or worry, often disproportionate to the actual threat.
2. **Generalized Anxiety Disorder (GAD):** It is marked by chronic, excessive worry about various aspects of life, such as work, health, or personal relationships. Symptoms include restlessness, fatigue, difficulty concentrating, irritability, muscle tension, and sleep disturbances (Shalev, 2009). The exact cause of GAD is not fully understood, but it likely involves a combination of genetic, environmental, and psychological factors.
3. **Panic Disorder:** It is characterized by recurrent, unexpected panic attacks—sudden periods of intense fear or discomfort, often accompanied by physical symptoms such as heart palpitations, sweating, trembling, shortness of breath, and dizziness. Individuals with panic disorder often live in fear of having another attack, which can significantly impair their daily lives (Shalev, 2009).
4. **Social Anxiety Disorder:** It involves an intense fear of social situations where one might be judged, embarrassed, or scrutinized by others. This fear can lead to avoidance of social interactions, which can hinder personal and professional development. Physical symptoms may include blushing, sweating, trembling, and nausea (Shalev, 2009).
5. **Depression:** It is a common and serious mood disorder that affects how a person feels, thinks, and handles daily activities. It is more than just a bout of the blues; it is a persistent feeling of sadness or loss of interest that can lead to various emotional and physical problems.
6. **Major Depressive Disorder (MDD):** It is characterized by a combination of symptoms that interfere with a person's ability

to work, sleep, study, eat, and enjoy activities once pleasurable. Symptoms include persistent sadness, hopelessness, loss of interest in activities, changes in appetite or weight, sleep disturbances, fatigue, and thoughts of death or suicide. The exact cause of MDD is complex, involving genetic, biological, environmental, and psychological factors.

7. **Persistent Depressive Disorder (PDD):** Also known as dysthymia, is a chronic form of depression with symptoms that last for at least two years. While the symptoms may be less severe than those of MDD, they are long-lasting and can significantly impair daily functioning.
8. **Post-Traumatic Stress Disorder (PTSD):** It is a disorder that can develop after exposure to a traumatic event, such as combat, natural disasters, serious accidents, or assault. Symptoms include intrusive memories of the trauma, flashbacks, nightmares, severe anxiety, and uncontrollable thoughts about the event (Shalev, 2009). Individuals with PTSD may also experience avoidance behaviors, negative changes in thinking and mood, and hyperarousal symptoms such as being easily startled and having difficulty sleeping.
9. **Psychosomatic Disorders:** These are conditions where psychological stress manifests as physical symptoms. These symptoms are real and can affect any part of the body, but they lack a clear physical cause. Common psychosomatic symptoms include headaches, stomach aches, fatigue, and muscle pain (Shalev, 2009).

2.7.1. Causes and Risk Factors

Several factors can contribute to the development of stress-related disorders:

1. **Genetic Predisposition:** A family history of mental health disorders can increase the likelihood of developing similar conditions.
2. **Environmental Stressors:** Chronic stressors such as financial difficulties, relationship problems, or work-related stress can trigger or worsen these disorders.
3. **Trauma:** Experiencing or witnessing traumatic events can lead to PTSD and other stress-related conditions.
4. **Personality Traits:** Individuals with certain personality traits, such as perfectionism or high neuroticism, may be more susceptible to stress-related disorders.
5. **Biological Factors:** Imbalances in neurotransmitters and hormones can play a role in the development of these conditions (Shalev, 2009).

2.7.2. Symptoms and Diagnosis

The symptoms of stress-related disorders can vary widely depending on the specific condition. Common symptoms include:

- Persistent worry or fear;
- Sadness or hopelessness;
- Physical symptoms like headaches or stomach aches;
- Sleep disturbances;
- Difficulty concentrating;
- Avoidance of social situations or activities;
- Intrusive memories or flashbacks of traumatic events.

Diagnosis typically involves a thorough evaluation by a healthcare professional, including a detailed medical history, physical examination, and psychological assessment. Questionnaires and diagnostic tools, such as the Diagnostic and Statistical Manual of Mental Disorders (DSM-5), may also be used.

2.7.3. Treatment and Management

Effective treatment for stress-related disorders often involves a combination of therapies:

1. **Psychotherapy:** Cognitive-behavioral therapy (CBT) is widely used to treat anxiety and depression. It helps individuals identify and change negative thought patterns and behaviors. Exposure therapy is effective for PTSD, helping individuals confront and process traumatic memories.
2. **Medication:** Antidepressants, anti-anxiety medications, and mood stabilizers can help manage symptoms. Selective serotonin reuptake inhibitors (SSRIs) and serotonin-norepinephrine reuptake inhibitors (SNRIs) are commonly prescribed for anxiety and depression.
3. **Lifestyle Changes:** Regular exercise, a healthy diet, sufficient sleep, and stress-reduction techniques such as mindfulness and meditation can improve overall well-being and reduce symptoms.
4. **Support Systems:** Social support from family, friends, and support groups can provide emotional comfort and reduce feelings of isolation.
5. **Professional Help:** In severe cases, hospitalization or intensive outpatient programs may be necessary to provide a structured environment for recovery (Chu et al., 2024).

2.7.4. Prevention

The following methods can be used to prevent stress-related disorders as stated by McCraty & Atkinson (2012). This entails addressing risk factors and promoting resilience. Strategies include:

1. **Stress Management:** Learning effective stress management techniques, such as time management, relaxation exercises, and coping strategies, can reduce the impact of stress.
2. **Healthy Lifestyle:** Maintaining a balanced diet, regular exercise, and adequate sleep can improve overall mental health.
3. **Early Intervention:** Seeking help at the early signs of stress or mental health issues can prevent the development of more severe disorders.
4. **Building Resilience:** Developing strong social connections, a positive mindset, and effective problem-solving skills can enhance resilience to stress.

Stress-related disorders are complex conditions with a significant impact on individuals and society. Understanding their causes, symptoms, and treatment options is essential for effective management and prevention. With appropriate intervention and support, individuals can lead fulfilling lives despite these challenges. Promoting awareness and reducing stigma around mental health can further enhance efforts to address stress-related disorders and improve overall well-being (Tran et al., 2020).

2.8. PREVENTING STRESS-RELATED ILLNESSES

Stress, an unavoidable part of modern life, is a major contributor to numerous health issues. From mental health disorders like anxiety and depression to physical conditions such as hypertension, heart disease, and weakened immune function, the effects of chronic stress are far-reaching and significant. Preventing stress-related illnesses requires a multi-faceted approach that encompasses lifestyle changes, mental health support, workplace interventions, and societal shifts.

2.8.1. Stress and Its Impacts

Stress is the body's normal reaction to perceived dangers or difficulties. It can be helpful in brief spurts, giving you the drive and concentration you need to get over challenges. On the other hand, persistent and ongoing stress, or chronic stress, may be harmful. Stress chemicals like cortisol and adrenaline are released as a result, and these can eventually harm a number of body systems.

The physical effects of chronic stress include increased heart rate and blood pressure, weakened immune response, digestive problems, and sleep disturbances. Psychologically, stress can lead to anxiety, depression, irritability, and cognitive impairments. Socially, it can strain relationships and reduce overall quality of life ('Impact of Stress, Immunity, and Signals from Endocrine and Nervous System on Fascia,' 2021).

2.8.2. Lifestyle Changes to Prevent Stress

One of the most effective ways to prevent stress-related illnesses is through lifestyle changes. Regular physical activity, for instance, has been shown to reduce stress levels by releasing endorphins—natural mood lifters (Chu et al., 2024). Exercise also improves sleep, increases energy levels, and enhances overall well-being.

A balanced diet is equally crucial. Consuming a variety of nutrient-rich foods helps the body cope with stress. Omega-3 fatty acids, found in fish and flaxseed, can reduce inflammation and improve brain function. Similarly, foods rich in antioxidants, like fruits and vegetables, combat oxidative stress (Schneiderman et al., 2005).

Adequate sleep is vital for stress management (Chu et al., 2024). Poor sleep can exacerbate stress, creating a vicious cycle. Establishing a regular sleep routine, creating a restful environment, and avoiding stimulants like caffeine and electronic screens before bed can improve sleep quality.

Mindfulness and relaxation techniques such as meditation, yoga, and deep breathing exercises can also help manage stress. These practices reduce the body's stress response, promote relaxation, and improve emotional regulation.

2.8.3. Mental Health Support

Preventing stress-related illnesses also requires robust mental health support. Access to mental health services, such as counseling and therapy, can provide individuals with coping strategies and tools to manage stress effectively (Tran et al., 2020). Cognitive-behavioral therapy (CBT), for example, has been proven effective in reducing stress by changing negative thought patterns and behaviors.

Social support is another critical factor. Strong relationships with family, friends, and community provide emotional support and a sense of belonging, which can buffer against stress. Encouraging open communication and fostering supportive environments can make a significant difference.

2.8.4. Workplace Interventions

The studies conducted by Tran et al. (2020) also indicated that the workplace is a common source of stress for many people. Therefore, employers have a vital role in preventing stress-related illnesses. Implementing workplace wellness programs that promote physical activity, healthy eating, and mental health can significantly reduce stress levels among employees.

Flexible working arrangements, such as remote work options and flexible hours, can also alleviate stress by allowing employees to balance work and personal responsibilities more effectively. Encouraging regular breaks and providing spaces for relaxation can help employees recharge and reduce burnout.

Training managers and supervisors to recognize signs of stress and provide appropriate support can create a healthier work environment. Offering resources like stress management workshops and employee assistance programs can further support employees' well-being (Tran et al., 2020).

2.8.5. Societal Shifts

Addressing stress on a societal level involves creating environments that promote overall well-being. Policies that support work-life balance, such as parental leave and affordable childcare, can reduce stress for families. Access to affordable healthcare, including mental health services, ensures that individuals can seek help when needed.

Educational initiatives that teach stress management skills from a young age can equip individuals with the tools to handle stress throughout their lives. Schools can incorporate mindfulness and emotional regulation into their curricula, promoting a healthier future generation (Campbell & Ehlert, 2012).

2.8.6. The Role of Technology

Tran et al. (2020) also report that, while technology can be a source of stress, it can also be part of the solution. Mobile apps and online platforms offering guided meditations, stress management courses, and virtual counseling provide accessible resources for managing stress. However, it is crucial to balance technology use with offline activities to prevent digital burnout.

In summary, preventing stress-related illnesses requires a holistic approach that addresses the various sources and manifestations of stress. By making lifestyle changes, accessing mental health support, implementing workplace interventions, and fostering societal shifts, individuals and communities can reduce the burden of stress-related illnesses. Promoting a balanced and healthy life, with adequate support systems, can lead to a more resilient population capable of thriving despite the inevitable challenges of modern life.

CHAPTER 3

STRESS MANAGEMENT TECHNIQUES

Contents

Managing stress is essential to preserving general health and well-being, especially in the hectic and fast-paced world of today. Stress is characterized as a typical physiological and psychological reaction to difficult circumstances, but prolonged stress can cause major health problems such as depression, heart disease, and a compromised immune system (Sharma & Rush, 2014). Because stress has many facets, managing it requires an all-encompassing strategy that incorporates several strategies suited to certain situations and needs. In addition to reducing these negative consequences, effective stress management strategies improve an individual's capacity to handle stress in the future.

Many studies conducted in the past few years have highlighted how crucial it is to deal with stress in a proactive manner. Numerous behavioral, cognitive, and physiological techniques have been found by researchers to be highly effective in minimizing the negative effects of stress. These tactics stem from an understanding of the mechanics underlying the body's stress response, which include the release of cortisol, the principal stress hormone, following the activation of the hypothalamic-pituitary-adrenal (HPA) axis (Cohen et al., 2016). The need for efficient management techniques is highlighted by the fact that prolonged stimulation of the HPA axis can result in a variety of health issues.

Furthermore, stress is a major public health concern, according to the World Health Organization (WHO), as it is a factor in a number of non-communicable diseases and mental health disorders (Sharma & Rush, 2014). Stress is a ubiquitous part of modern life, therefore it's important for people to have useful, research-backed coping mechanisms at their disposal. In addition to individual efforts, this calls for the establishment of institutional and societal support networks that create atmospheres that are supportive of stress management and mental health.

This chapter focuses on stress management and discusses the following approaches to management.

3.1. RELAXATION TECHNIQUES

Effective relaxation methods that ease both physiological and psychological tension are necessary for managing chronic stress. These are five well-known methods of relaxation:

3.1.1. Deep Breathing Exercises

Due to their non-pharmacological approach to improving both physical and psychological well-being, deep breathing exercises are essential for controlling chronic stress. The parasympathetic nervous system, which opposes the body's stress reaction normally controlled by the sympathetic nervous system, and is primarily stimulated by deep breathing. Reduced heart rate, blood pressure, and cortisol levels are characteristics of this physiological change that all work together to promote relaxation (Sharma & Rush, 2014).

Deep breathing techniques have been shown to be effective in lowering long-term stress. A 2017 study by Ma et al. showed that people who used deep breathing exercises reported feeling much less stressed and had better moods overall. To objectively evaluate the effect of the intervention, the study used measurements like the salivary cortisol levels and the Perceived Stress Scale (PSS). The results demonstrated the possibility of deep breathing as an effective stress management technique by showing a strong link between it and decreased physiological indicators of stress.

Moreover, studies have demonstrated that deep breathing exercises enhance heart rate variability (HRV), a measure of the autonomic nervous system's adaptability and stress tolerance (Ma et al., 2017). Improved cardiovascular health and a more flexible stress response are linked to higher HRV. This implies that deep breathing exercises on a regular basis can improve the body's capacity to handle stressors more effectively.

Deep breathing exercises are a popular choice for stress treatment because of their accessibility and simplicity. They are especially helpful for those with hectic schedules because they don't require specialized equipment or a lot of training, and they can be done in a variety of environments. Furthermore, incorporating deep breathing exercises into regular activities can act as a prophylactic strategy, thereby slowing down the progression of stress-related illnesses over time (Hasan & Tumah, 2019).

3.1.2. Progressive Muscle Relaxation (PMR)

A well-known method for treating chronic stress is progressive muscle relaxation (PMR), which has major physiological and psychological advantages. PMR, which was created by Edmund Jacobson in the early 20th century, entails methodically tensing and then relaxing various body muscular groups to induce a physically relaxed state that can reduce stress (Kaur, 2016). This approach is based on the idea that people can lower their overall stress levels by learning to intentionally relax their tense muscles, which are a common reaction to stress.

Studies have consistently shown that PMR is a useful tool for managing stress. According to a meta-analysis by Glazer & Liu (2017), which examined numerous studies on PMR, it considerably lowers tension in the muscles, stress, and anxiety. According to the study, PMR can lower cortisol levels, blood pressure, and heart rate by inhibiting the sympathetic nervous system, which is in charge of the body's "fight or flight" reaction. For people who are under constant stress, this physiological change is essential since prolonged sympathetic nervous system activation can result in a variety of health problems, including hypertension and cardiovascular disease.

Additionally, PMR has been demonstrated to enhance mental health results. According to Kaur (2016), PMR practitioners significantly reduced their symptoms of anxiety and depression. According to the study, PMR can improve bodily awareness and mindfulness, which can improve emotional control and stress tolerance.

PMR is a useful technique for stress management because of its accessibility and simplicity of use. Being able to conduct it in many locations and not requiring any additional equipment makes it especially useful for people with hectic schedules. Long-term gains in stress resilience and general well-being can result from consistent PMR practice (Glazer & Liu, 2017) (Figure 3.1).

Figure 3.1. *Benefits of progressive muscle relaxation.*

Source: Kevin-Cleary (2023).

3.1.3. Guided Imagery

Using the power of imagination to encourage relaxation and lower stress levels, guided imagery is a useful method for managing chronic stress. This technique uses numerous senses to produce a realistic mental experience while encouraging people to visualize serene and relaxing images or scenarios. The parasympathetic nerve system, which counteracts the body's stress reaction and lowers blood pressure, heart rate, and stress hormone levels, is activated by guided imagery (Glazer & Liu, 2017).

The effectiveness of guided imagery in stress reduction is supported by research. In a review of several relaxation methods, Glazer & Liu (2017) discovered that guided imagery dramatically lowered participants' levels of stress and anxiety. According to the study, guided imagery promotes relaxation and well-being by giving people a mental getaway and a way to divert their attention from troubling thoughts. Furthermore, studies have conncted guided imagcry to better mental health outcomes, including fewer symptoms of anxiety and sadness and increased coping mechanisms (Sharma & Rush, 2014).

Guided imagery is an effective method for stress management since it is easily accessible. It doesn't require any specific equipment to practice, and it may be done on your own or with recordings. Its adaptability makes it simple to incorporate into regular activities, which makes it a useful tactic for lowering long-term stress and enhancing mental wellness in general (Glazer & Liu, 2017).

3.1.4. Biofeedback

Biofeedback is a well-known method for treating chronic stress. It uses real-time data to assist people in taking control of physiological processes like heart rate, muscular tension, and skin temperature that are usually thought to be automatic. Biofeedback gives people quick feedback by utilizing electronics to track these physiological signals, enabling them to change how their bodies react to stress (Glazer & Liu, 2017).

The usefulness of biofeedback in lowering chronic stress has been repeatedly shown by research. Biofeedback dramatically lowered stress-related symptoms, such as anxiety and hypertension, according to a meta-analysis conducted in 2014 by Schoenberg and David. The results of the study demonstrated how biofeedback supports physiological balance and relaxation by raising people's awareness of stress reactions and teaching them how to actively manage them. Of particular interest is the capacity of biofeedback to elevate heart rate variability (HRV), a crucial measure of autonomic nervous system function. Better cardiovascular health and stress resilience are linked to higher HRV (Hasan & Tumah, 2019). Furthermore, it has been demonstrated that biofeedback enhances mental health results. Studies show that in people with chronic stress, biofeedback therapies significantly reduce anxiety and enhance coping strategies (Glazer & Liu, 2017; Hasan & Tumah, 2019). This implies that biofeedback promotes long-term stress management abilities in addition to reducing acute stress symptoms.

3.1.5. Autogenic Training

One effective self-relaxation method for handling chronic stress is autogenic training (AT). The early 20th-century pioneer Johannes Schultz developed acceptance therapy (AT), which involves repeating a series of self-statements about feeling warm and heavy in various body parts to produce deep relaxation and lessen stress reactions (Sharma & Rush, 2014).

Studies with empirical data have shown how effective AT is at managing stress. It dramatically lessens depressive, anxious, and stressed-out symptoms. According to Glazer & Liu's (2017) analysis, AT lowers blood pressure, heart rate, and cortisol levels by modifying the autonomic nervous system, reducing sympathetic activity, and increasing parasympathetic responses.

Moreover, evidence suggests that AT enhances mental health results. According to research, people who practice AT reported much higher levels of psychological well-being and lower levels of perceived stress. It is believed that AT works by refocusing cognitive attention from stressful thoughts to peaceful, body-focused experiences, which promotes mindfulness and relaxation (Sharma & Rush, 2014).

Due to its ease of use and accessibility, AT is a useful tool for stress management. Because it can be practiced anywhere and doesn't require any specific equipment, people may easily include it in their everyday routines. Because of its adaptability, AT is a useful tactic for improving general mental health and managing long-term stress (Glazer & Liu, 2017).

3.2. MINDFULNESS AND MEDITATION

There is strong empirical evidence to support the widespread recognition of mindfulness and meditation's effectiveness in stress management. These techniques focus on developing a nonjudgmental acceptance of one's ideas, feelings, and physical experiences as well as present-moment awareness. The fundamental idea is to divert attention from thoughts that cause stress and instead focus on being in a calm awareness state, which has several physiological and psychological advantages (Sharma & Rush, 2014).

One of the most researched mindfulness initiatives is Jon Kabat-Zinn's mindfulness-based stress reduction (MBSR). In order to improve awareness and acceptance of the present moment, MBSR combines mindfulness meditation with yoga (Tran et al., 2020). Studies have shown that MBSR considerably lowers anxiety, sadness, and stress. According to a meta-analysis conducted by Khoury et al. (2015), MBSR treatments significantly lower stress levels in a variety of demographics, improving mental health and general well-being.

Mindfulness and meditation techniques are physiologically linked to decreased activity in the hypothalamic-pituitary axis (HPA axis), which controls the body's stress response. According to studies, these techniques promote relaxation by lowering cortisol levels, blood pressure, and heart rate (Glazer &

Liu, 2017). Studies on neuroimaging also show that mindfulness and meditation improve the prefrontal cortex and anterior cingulate cortex, two brain regions involved in regulating emotions, while at the same time lowering activity in the amygdala, a region linked to the stress response (Sharma & Rush, 2014).

Additionally, practicing mindfulness and meditation increases cognitive flexibility and resilience, which helps people deal with challenges more skillfully. According to Tran et al. (2020), they improve attentional control, enabling people to break out from negative and ruminative thought patterns that worsen stress. These techniques support emotional control and a balanced viewpoint, which help with long-term stress reduction and mental wellness.

Managing chronic stress can be effectively achieved through various mindfulness and meditation-based practices. Here are the widely recognized techniques:

3.2.1. Mindfulness-Based Stress Reduction (MBSR)

A systematic approach called Mindfulness-Based Stress Reduction (MBSR), created by Jon Kabat-Zinn, is very helpful in treating long-term stress. Over the course of eight weeks, it integrates body scanning, mild yoga, and mindfulness meditation to foster a deeper understanding and acceptance of the present moment (Sharma & Rush, 2014). Developing mindfulness skills that people may use in stressful situations is the main goal of Mindfulness-Based Stress Reduction (MBSR), which helps people lessen the negative effects of chronic stress on their bodies and minds (Figure 3.2).

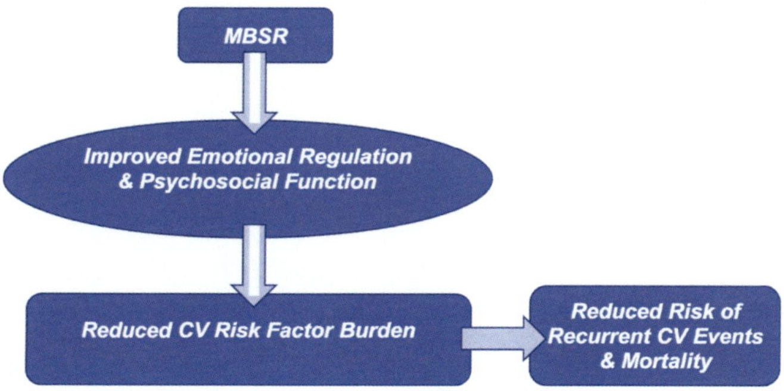

Figure 3.2. *Conceptual model of MBSR in cardiac patients.*

Source: Nijjar et al. (2019).

Empirical studies substantiate MBSR's effectiveness in stress reduction. After reviewing several studies, Khoury et al. (2015) conducted a meta-analysis and discovered that MBSR significantly lowers stress, anxiety, and depression.

The main stress hormone, cortisol, is found at lower levels in MBSR program participants, which suggests a decrease in physiological stress reactions. Furthermore, research using neuroimaging has demonstrated that MBSR can result in anatomical alterations in the brain, including an increase in gray matter density in areas of the brain linked to emotion regulation, like the hippocampus and prefrontal cortex (Sharma & Rush, 2014).

Additionally, it has been demonstrated that MBSR enhances emotional control and cognitive flexibility, which leads to better mental health outcomes. Through cultivating an impartial consciousness of ideas and emotions, people get the ability to react to pressures in a more adaptive way as opposed to a reactive one. In addition to promoting long-term resilience and well-being, this lowers the overall experience of stress (Sharma & Rush, 2014).

3.2.2. Body Scan Meditation

Body scan meditation is a mindfulness practice that significantly contributes to managing chronic stress by promoting heightened body awareness and relaxation. This technique involves systematically focusing attention on different parts of the body, usually starting from the toes and moving upward, noticing sensations without judgment. The practice aims to cultivate a deep sense of relaxation and awareness of the present moment, which can alleviate stress (Khoury et al., 2015).

Empirical evidence supports the effectiveness of body scan meditation in reducing chronic stress. Participants in an eight-week mindfulness-based stress reduction (MBSR) program, which includes body scan meditation, experienced significant reductions in perceived stress and improvements in psychological well-being (Hasan & Tumah, 2019). The study highlighted that body scan meditation helps individuals become more aware of bodily sensations, which can lead to early recognition of stress responses and more effective stress management.

Moreover, body scan meditation has been shown to decrease physiological markers of stress. Regular practice of body scan meditation lowers heart rate and cortisol levels, indicating reduced physiological arousal and a shift towards parasympathetic nervous system dominance. This shift promotes relaxation and recovery from stress (Glazer & Liu, 2017).

Additionally, body scan meditation enhances emotional regulation and resilience. By encouraging a non-judgmental awareness of bodily sensations, practitioners can develop greater acceptance and reduce the impact of negative emotions associated with chronic stress (Hasan & Tumah, 2019).

3.2.3. Breath Awareness Meditation

By encouraging a focused, serene, and present-centered awareness, breath awareness meditation is a mindfulness technique that is extremely helpful in reducing chronic stress. This practice entails focusing on the breath's inherent rhythm, feeling the inhalation and exhalation, and gently bringing one's attention back to the breath whenever the mind strays (Glazer & Liu, 2017).

The effectiveness of breath awareness meditation in lowering chronic stress is supported by empirical research. Breath-focused meditation, even for short periods of time, can dramatically improve mood and reduce stress. By encouraging a state of relaxation and mental clarity, this technique aids in disengaging from stressors and ruminative thoughts (Tran et al., 2020).

From a physiological perspective, breath awareness meditation modifies the autonomic nerve system to increase parasympathetic activity, which balances the stress response in the body. Breath awareness involves deep, slow breathing that lowers blood pressure, heart rate, and cortisol levels while raising heart rate variability (HRV), which suggests better autonomic regulation and stress resilience (Glazer & Liu, 2017).

Additionally, breath awareness meditation improves cognitive performance and emotional control. People who keep their attention on their breathing become more attentive, which can enhance their coping skills and lessen the negative psychological effects of stress (Sharma & Rush, 2014). Additionally, by promoting focus and mental clarity, this practice helps lessen the cognitive exhaustion that is frequently caused by chronic stress.

Finally, it should be noted that breath awareness meditation is a tried-and-true method of treating chronic stress. It is an effective means of promoting general mental and physical well-being because it facilitates emotional management, increases physiological relaxation, and improves cognitive performance.

3.3. EXERCISE AND STRESS REDUCTION

Exercise is essential for stress management since it has positive effects on the body and the mind. Frequent exercise is proven to lower stress levels via a number of processes, offering a comprehensive strategy to improve general well-being. Exercise physiologically lowers stress by adjusting the body's stress response mechanisms. Physical activity increases the creation of endorphins, which are naturally occurring mood enhancers, and simultaneously lowers levels of cortisol, the main stress hormone (Kaur, 2016). This physiological change encourages relaxation and a sense of well-being. Moreover, cardiovascular health and system efficiency are improved by exercise, which can lessen the negative impacts of stress on the body. Engaging in physical activity improves mood and cognitive function, which helps people psychologically manage stress. Regular exercise has been associated with a significant reduction in symptoms of depression and anxiety. Exercise reduces stress from daily life and increases feelings of self-efficacy and accomplishment. Furthermore, it has been shown that exercise boosts cognitive resilience and neuroplasticity, which fortifies the brain's capacity to adapt to stress (Kaur, 2016).

Regular exercise also improves the quality of sleep, which is crucial for lowering stress. According to Glazer & Liu (2017), while having enough sleep enhances the body's ability to manage stress, obtaining too little sleep may actually make stress worse. Because exercise improves sleep patterns, it indirectly reduces stress and its effects. Some ways through which exercise helps in mitigating chronic stress are discussed in subsections.

3.3.1. Reduction of Stress Hormones and Production of Endorphins

By reducing the synthesis of stress hormones and raising endorphin levels, exercise is essential for controlling chronic stress. The main stress hormone generated by the hypothalamic-pituitary-adrenal (HPA) axis, cortisol, is usually raised in response to chronic stress. Engaging in regular physical activity can help to reduce this reaction. Regular exercisers have reduced cortisol responses to acute stress, which suggests a more controlled HPA axis and a less physiological stress response (Sharon-David & Tenenbaum, 2017).

Additionally, endorphins are neurochemicals released by the pituitary and central nervous systems in response to exercise. Because these endogenous opioids function as organic analgesics and mood enhancers, they are frequently referred to as "feel-good" hormones. The "runner's high," a condition of euphoria and decreased sense of pain, is a result of the rise in endorphin levels that occurs after physical activity (Sharon-David & Tenenbaum, 2017). This physiological reaction not only elevates mood but also mitigates the adverse impacts of stress by encouraging a feeling of wellness and calmness.

Regular exercise also enhances general autonomic function and stress resilience by improving the balance between sympathetic and parasympathetic nervous system activity (Sharon-David & Tenenbaum, 2017). This physiological adjustment is essential for lessening the long-term effects of persistent stress on the body, which includes a decreased likelihood of disorders linked to stress.

3.3.2. Improvement in Cardiovascular and Overall Physical Health

Cardiovascular and general physical health are greatly enhanced by exercise, and these benefits are critical in the management of chronic stress. By raising cardiac output, enhancing endothelial function, and encouraging improved lipid profiles, regular physical activity improves cardiovascular health. According to Kivimäki & Steptoe (2018), these adaptations lower the risk of cardiovascular illnesses, which are frequently made worse by prolonged stress.

Lower blood pressure and improved arterial stiffness are two important ways that exercise promotes cardiovascular health. By lowering blood pressure at both the systolic and diastolic levels, aerobic exercise improves cardiovascular health in general. Enhancing overall physical health and stress resilience, improved vascular health promotes more efficient circulation that enables better oxygen and nutrition delivery to tissues (Kivimäki & Steptoe, 2018).

Exercise also helps with weight control and belly fat reduction, both of which are associated with decreased inflammatory levels and better metabolic health. This is significant because, according to Kivimäki & Steptoe (2018), prolonged stress is linked to an increase in abdominal obesity and metabolic syndrome. Frequent exercise lowers the physiological toll of stress and promotes a healthier body composition, both of which contribute to offset these risks.

Exercise also strengthens the immune system, which lowers the risk of disorders linked to stress. Exercise that is moderate in intensity improves immune surveillance and lowers the risk of upper respiratory tract infections, which are often made worse by prolonged stress (Kivimäki & Steptoe, 2018).

Along with lowering inflammation and improving metabolic health, exercise also aids in weight loss and the elimination of abdominal fat. This is noteworthy because extended stress is associated with an increase in metabolic syndrome and abdominal obesity, according Kivimäki & Steptoe (2018). Regular exercise helps to mitigate these risks by promoting a healthy body composition and reducing the physiological toll that stress takes.

Additionally, exercise fortifies the immune system, reducing the likelihood of illnesses associated with stress. Moderately intense exercise strengthens immune surveillance and reduces the incidence of upper respiratory tract infections, which are frequently exacerbated by extended periods of stress (Kivimäki & Steptoe, 2018).

3.4. HEALTHY EATING HABITS

Research indicates that many people alter their eating habits while under a lot of stress. Increasing caffeine intake to boost productivity in the morning or at night is one of the most popular dietary modifications linked to stress. Ironically, coffee might worsen the stress response even though it is linked to brief gains in alertness. For instance, caffeine causes the release of catecholamines such as dopamine, norepinephrine, and epinephrine (adrenaline), as well as several stress hormones such as cortisol and glucocorticoids. Hours after consumption, the production of these compounds is linked to elevated stress levels (Chu et al., 2024). In addition, caffeine consumption can lead to other conditions that can affect the body's ability to respond to stress, including insomnia, hypertension, an increased risk of heart disease, gastrointestinal problems, and immune system suppression, making you more prone to infections.

As noted earlier, high levels of stress are also associated with increases in cortisol, which tend to lead to cravings for high-fat or sugary foods (Tran et al., 2020). In addition, when busy, people often resort to skipping meals or eating fast foods. These dietary changes can make it more difficult for our bodies to manage stress. Eating a healthy diet full of nutrient-rich foods can help you manage stress in several ways. For example, complex carbohydrates, like oatmeal and whole grain breads and cereals, cause your brain to release Serotonin, a neurotransmitter associated with positive mood. Even simple carbohydrates, such as candy or chocolate can be good for a quick spike in Serotonin. Research suggests that foods rich in Vitamin C (e.g., oranges) and Omega-3 fatty acids (e.g., salmon) can help reduce levels of stress hormones and improve immune functioning. A well-balanced diet improves one's ability to manage stress (Hill et al., 2022).

3.4.1. Nutrient Balance and Brain Health

A diet that is well-balanced guarantees that the body gets the necessary nutrients, like vitamins, minerals, and antioxidants, which are critical for the best possible brain function. A few key nutrients are important for stress reduction and mental wellness maintenance. B vitamins, for example, B6, B12, and folate, are essential for the synthesis of neurotransmitters that control mood and emotional reactions, like dopamine and serotonin. Whole grains, lean meats, eggs, dairy products, leafy green vegetables, beans, and legumes are foods high in B vitamins (Hill et al., 2022).

Another essential element for brain health is omega-3 fatty acids, which are included in fatty fish like salmon, mackerel, and sardines as well as flaxseeds, walnuts, and chia seeds. These fatty acids have anti-inflammatory properties and have been shown to reduce symptoms of anxiety and depression (Yaribeygi et al., 2017). They help maintain the integrity of brain cell membranes and support the transmission of neural signals, which can enhance mood and cognitive function.

3.4.2. Stable Blood Sugar Levels

Stress management requires steady blood sugar levels to be maintained. Significant blood sugar changes can cause weariness, mood swings, and irritation, all of which can make stress worse. Blood sugar levels can be maintained by eating regular, well-balanced meals that contain a variety of complex carbs, proteins, and healthy fats.

Whole grains, legumes, vegetables, and fruits are examples of complex carbs that release glucose into the bloodstream gradually, avoiding sharp rises and falls in blood sugar levels. Lean meats, fish, eggs, dairy products, beans, and nuts are examples of foods high in protein that assist in maintaining energy levels and prolong satiety. Healthy fats, including those found in avocados, nuts, seeds, and olive oil, also contribute to stable blood sugar levels by slowing the digestion and absorption of carbohydrates.

3.4.3. Gut-Brain Linkage and Gut Health

The gut-brain axis, which describes the relationship between gut health and mental health, is being highlighted by new studies. The population of microbes that reside in the digestive system, known as the gut microbiome, is essential for controlling mood and stress reactions (Yaribeygi et al., 2017). Thus, a diet that supports a balanced gut microbiota may help with stress management.

Fruits, vegetables, whole grains, and legumes are examples of foods high in fiber that promote the development of good gut bacteria. Probiotics are good bacteria that are added to the gut and can be found in fermented foods such as kefir, yogurt, sauerkraut, kimchi, and kombucha. Prebiotics, which are non-digestible fibers that feed beneficial gut bacteria, are found in foods like garlic, onions, leeks, asparagus, bananas, and oats (Hill et al., 2022). By promoting a healthy gut microbiome, these dietary components can influence the production of neurotransmitters and other signaling molecules that affect mood and stress levels.

3.4.4. Hydration and Mental Ability

Although it is frequently disregarded, adequate hydration is crucial for mood management and cognitive performance. Headaches, exhaustion, and trouble focusing can all result from even slight dehydration, and these symptoms can raise stress levels. To stay hydrated and promote general health, it's critical to drink enough water throughout the day. Although the precise quantity of water required can change based on age, gender, activity level, and environment, it's generally recommended to drink at least eight 8-ounce glasses of water each day.

3.4.5. Cutting Sugar and Caffeine Intake

Another key component of practicing good eating practices to reduce stress is limiting sugar and caffeine intake. Although caffeine might give you a short-term energy boost, too much of it can make you jittery, raise your heart rate, and make you more sensitive to stress (Tran et al., 2020). Additionally, it may disrupt sleep, which is essential for stress reduction. Cutting less on coffee, especially in the afternoon and evening, can improve sleep quality and lessen stress.

Similarly, consuming too much sugar can cause mood swings and energy dips, both of which can make stress worse. Sugar-filled foods and drinks quickly raise blood sugar levels, which are then sharply lowered, which can have an impact on mood and energy (Hill et al., 2022). Reducing the consumption of sugary foods and beverages and opting for whole, unprocessed foods can help maintain more stable energy levels and moods.

3.4.6. Eating with Awareness and Unwinding

The practice of mindful eating involves paying close attention to all aspects of the eating experience, such as the food's flavor, texture, and aroma as well as your body's signals of hunger and fullness. Slower eating, improved digestion, and a deeper appreciation of food are all encouraged by this technique. Mindful eating can help people unwind and feel less stressed since it encourages them to pay attention to the present and completely engage with the meal (Hill et al., 2022).

Reducing emotional eating, improving one's connection with food, and making more deliberate food choices are all benefits of mindful eating. The mindful eating experience can be improved by using practices like inhaling deeply before eating, enjoying every bite, and avoiding electronics like televisions and cellphones during meals.

Eating a healthy diet can be very helpful in reducing stress and enhancing general well-being. Both physical and mental health can be supported by a balanced diet that contains vital nutrients, stable blood sugar levels, a healthy gut microbiota, enough hydration, and a reduction in the consumption of sugar and caffeine. Furthermore, mindful eating can ease tension and encourage relaxation. People can improve their overall quality of life and their capacity to handle stress by adopting these healthy eating habits into their daily lives.

3.5. TIME MANAGEMENT STRATEGIES

Finding the time to practice stress management techniques is one of the first issues people run into. Whether it's reading a chapter, working on a homework assignment, or using a relaxation technique. Not only is time management a stress management topic, but it is also a need for all other work-related stress management strategies (Shindo & Cohen, 1976). People frequently complain about not having enough time or feeling under pressure from it, and they view time as a cause of stress. It is incorrect, though, to consider time as a stressor because there are only 24 hours in a day, regardless of any method. When people have concerns about time, they usually have problems with too much work, not enough recreation, conflicts between equally unpleasant (or unpleasant) tasks, missed deadlines, feeling overwhelmed by trivia, inability to do what they want because of unfinished tasks, and so on.

3.5.1. Understanding Time Management and Stress

Planning and regulating how much time is spent on various tasks is part of time management, particularly when trying to boost productivity, effectiveness, and efficiency. Contrarily, stress is a physical, mental, or emotional condition that results in tension in the body or mind. Stress and time management go hand in hand; although ineffective time management can make stress worse, competent time management can help reduce it.

3.5.2. Key Time Management Strategies

1. **Prioritization:** One of the most effective time management strategies is prioritization. Prioritizing tasks helps focus on what is most important and urgent, ensuring that critical activities are completed on time. The Eisenhower Matrix is a popular tool for prioritization (Smith, 2002).

It categorizes tasks into four quadrants:

- Urgent and important: Do these tasks immediately;
- Important but not urgent: Schedule these tasks for later;
- Urgent but not important: Delegate these tasks if possible; and
- Neither urgent nor important: Eliminate these tasks.

By focusing on important tasks first, you can prevent last-minute rushes and reduce stress.

2. **SMART Goal-Setting:** Achievable goals must be set in order to manage time effectively. Goals are made sure to be clear and reachable by the SMART criteria, which stand for Specific, Measurable, Achievable, Relevant, and Time-bound. SMART goals lower anxiety and increase motivation by giving a clear path and a sense of accomplishment when reached (Rao, 2014).
3. **Making a Timetable:** A well-organized schedule is the foundation of effective time management. A planner or digital calendar can be used to set aside certain times for various tasks. Divide your day into periods that are allocated to different kinds of work or chores

(Rao, 2014). This lessens stress by keeping projects on track and preventing them from taking up time that should be spent on other activities.

4. **Dividing Work into Manageable Steps:** Big jobs can be difficult and overpowering. They become less intimidating and simpler to handle when broken down into smaller, more doable tasks. This method, known as "chunking," lets you go forward steadily and gives you lots of chances to enjoy little victories, which can lift your spirits and ease tension (Rao, 2014).
5. **Blocking Out Time:** Time blocking is breaking up your day into intervals of time that are allotted to a certain task or set of related chores. You may focus on one task at a time by doing this, which improves productivity and focus. Additionally, it facilitates the establishment of boundaries between work and personal time, improves work-life balance, and lowers stress (Rao, 2014).
6. **Steer Clear of Multitasking:** Although multitasking may seem like a good method to accomplish more in less time, it frequently results in lower productivity and higher levels of stress. Your brain must continually move between tasks when you multitask, which increases the likelihood of errors and lengthens the time it takes to finish tasks. Concentrating on a single job at a time can boost productivity and lessen tension.
7. **Taking Vacancies Often:** Regular breaks are crucial for stress reduction and productivity maintenance. Methods that help sustain attention and avert burnout include the Pomodoro Technique, which entails working for a certain amount of time—typically 25 minutes—followed by a brief break. Taking regular pauses helps your brain recover and function better overall.
8. **Assigning Assignments:** One crucial aspect of time management is delegation. Assign duties to others if you have any that they can do. This lessens your workload and frees up time for more important tasks, which lowers stress levels. Clear communication and having faith in others to do the job are essential components of effective delegation.
9. **Reducing Interruptions:** Distractions are a big source of stress and time wastage. Finding and removing distractions from your surroundings is essential to efficient time management. This could entail establishing limits with coworkers, disabling pointless notifications, or setting up a special area for work. Reducing

distractions improves your ability to concentrate and finish work faster.

10. **Taking Care of Oneself:** One crucial but frequently disregarded aspect of time management is self-care. Make sure you maintain a nutritious diet, get adequate sleep, and exercise frequently. Maintaining your mental and physical well-being raises your general quality of life and sharpens your time management skills. Self-care techniques can also lessen stress by assisting you in maintaining composure and attention.
11. **Examining and Contemplating:** Continuous improvement requires regular assessment and reflection on your time management techniques. Evaluate the successes and failures of each day or week at the end. Adapt your methods as necessary to increase productivity. You may better manage your time and stress levels by learning from your experiences and making the appropriate adjustments through reflection.

3.5.3. Putting Time Management Techniques into Practice

To put these time management techniques into practice, dedication, and regularity are needed. Determine which aspects of your life you find most difficult to manage your time effectively, then progressively use the techniques that work best for you. To get you started, follow these steps:

- i. **Identify Your Priorities:** Determine what is most important to you and focus on those areas.
- ii. **Set Clear Goals:** Use the SMART criteria to set clear, achievable goals.
- iii. **Create a Schedule:** Plan your day in advance, allocating specific time slots for different tasks.
- iv. **Break Down Tasks:** Divide large tasks into smaller steps to make them more manageable.
- v. **Use Time Blocks:** Dedicate blocks of time for specific activities to maintain focus.
- vi. **Limit Multitasking:** Focus on one task at a time to improve efficiency.
- vii. **Take Breaks:** Incorporate regular breaks into your schedule to prevent burnout.
- viii. **Delegate When Possible:** Don’t hesitate to delegate tasks to others.

ix. **Minimize Distractions:** Identify and reduce distractions in your environment.

x. **Practice Self-Care:** Take care of your physical and mental health to improve overall well-being.

xi. **Reflect and Adjust:** Regularly review your time management strategies and make necessary adjustments.

One of the most effective tools for controlling stress is effective time management. You can greatly reduce stress and increase productivity by prioritizing tasks, setting SMART goals, making a schedule, breaking down tasks, using time blocks, avoiding multitasking, taking regular breaks, delegating tasks, limiting distractions, engaging in self-care, and routinely reviewing and reflecting on your strategies. Although putting these tactics into practice will take time and work, the rewards in terms of improved mental health and general quality of life will be well worth it. Develop habits that will help you better manage your time and stress by starting small, being patient with yourself, and building them gradually.

3.6. SOCIAL SUPPORT AND STRESS

Human existence is filled with stress, which can take many different forms and have varying degrees of intensity. It is a result of ongoing demands, traumatic experiences, life transformations, and daily obstacles. Even though stress is an unavoidable part of life, the availability of social support can have a big impact on how people view and manage stress. The emotional, practical, informational, and appraisal resources that one's social network—which might include friends, family, coworkers, and neighbors—offers are referred to as social support (Baum, 1984). This paper examines the complex relationship between stress and social support, emphasizing how social support reduces stress and its effects on mental and physical health.

3.6.1. Social Support Types

There are four primary categories of social support as stated by Glanz et al. (2008):

1. **Emotional Support:** This entails showing compassion, love, trust, and empathy. People who receive emotional support are more likely to feel understood and respected, which helps lessen feelings of worry and loneliness.
2. **Instrumental Support:** This refers to material aid like cash, products, or services. For example, a friend offering to help with

housework or a family offering financial support can ease stress by taking on less practical responsibilities.

3. **Informational Support:** This includes information, recommendations, and guidance that can assist people in coping with pressures. People can manage stressful situations more skillfully if they are given advice on coping mechanisms or knowledge about the tools that are accessible.
4. **Appraisal Support:** This type of assistance helps people evaluate their stressors and coping strategies by providing them with feedback and affirmation. Positive feedback and constructive criticism can boost self-esteem and provide a sense of control over stressful situations (Glanz et al., 2008).

3.6.2. The Impact of Social Support on Stress

Social support and stress have a complex interaction that involves a number of psychological and physiological mechanisms (Ozbay et al., 2007). Stress can be lessened by social support via the following channels:

1. **Perceived Availability:** Just having the conviction that help is out there might reduce stress. It can be consoling and lessen the sense of how bad tension is when you know that there are people out there who are willing and able to assist.
2. **Stress Appraisal:** Perceptions of stressors can be influenced by social support. Stressors can be reframed more positively through supportive relationships, appearing less dangerous and more manageable.
3. **Emotional Regulation:** Emotional support helps in the regulation of negative emotions. Talking to someone who listens and empathizes can lead to a release of pent-up emotions, reducing emotional burden and promoting psychological well-being.
4. **Physiological Effects:** Social support can have direct physiological benefits, such as lowering blood pressure and reducing levels of stress hormones like cortisol. Positive social interactions activate the body's parasympathetic nervous system, promoting relaxation and recovery.
5. **Behavioral Changes:** Supportive relationships can encourage healthy behaviors and discourage harmful ones. For example, friends and family can motivate individuals to engage in physical activities, eat healthily, or avoid substance abuse, all of which contribute to better stress management (Ozbay et al., 2007).

3.6.2.1. Evidence From Research

The positive impacts of social support on stress and health outcomes have been shown in numerous studies. For example, studies have indicated that those who have supportive social networks have lower rates of anxiety and depression. Social support has direct and buffering benefits on mental health, lessening the negative impacts of stresses on psychological well-being, according to a study by Cohen & Wills (1985).

A seminal study conducted over nine years in 1979 by Berkman and Syme found that those with higher levels of social integration had lower death rates in the context of physical health. This research emphasizes how important social support is for long-term health and well-being.

3.6.3. Social Support in Different Contexts

The role of social support varies across different life contexts and populations. For instance, in the workplace, social support from colleagues and supervisors can enhance job satisfaction and reduce job-related stress (Smith, 2002). In academic settings, support from peers and mentors can improve academic performance and reduce anxiety among students.

In healthcare, social support is crucial for patients dealing with chronic illnesses. Support groups, counseling, and family involvement can improve patients' adherence to treatment regimens and enhance their quality of life.

3.6.4. Obstacles to Efficient Social Assistance

Not everyone has access to sufficient social support, despite its advantages. Several things can make assistance less accessible and ineffective, including:

Social isolation: A major obstacle to receiving social support is loneliness and social isolation. People who live in rural places, have impairments or are elderly are especially susceptible to isolation.

Stigma and shame: Social and cultural standards may attach a stigma to asking for assistance. People may be reluctant to ask for help because they are humiliated or afraid of being judged.

Quality of support: Not all support is beneficial. Negative interactions, such as criticism or excessive control, can exacerbate stress rather than alleviate it. The quality and nature of support are crucial for its effectiveness.

Personal factors: Personality traits and individual differences can influence the ability to seek and receive support. For example, introverted individuals may find it challenging to ask for help, while those with low self-esteem may doubt the availability of support (Rickwood et al., 2007).

3.6.5. Consequences of Intervention

Designing interventions targeted at enhancing mental and physical health requires an understanding of the connection between stress and social support. Among the techniques suggested by Ozbay et al. (2007) to strengthen social support are:

1. **Community Building:** Encouraging social media and community involvement can help lessen social isolation and improve access to resources.
2. **Help Groups:** Targeted help can be given by creating support groups for particular demographics, such as those going through major life transitions or people with chronic illnesses.
3. **Education and Training:** People can be empowered to create and preserve supportive connections by learning the value of social support and receiving training in effective communication and support-seeking techniques.
4. **Policy Initiatives:** Implementing policies that promote social welfare, such as affordable healthcare, mental health services, and programs for the elderly, can enhance the overall availability of social support (Ozbay et al., 2007).

3.7. SEEKING PROFESSIONAL HELP FOR STRESS

Life is full of stress, which affects people from all backgrounds and demographics. Stress can be a motivator in moderation, but prolonged or severe stress can be harmful to one's physical and emotional well-being. Effective stress management strategies are crucial because of the widespread nature of stress and its potential effects. Securing expert assistance is a crucial tactic that is frequently overlooked (Hubbard et al., 2018) (Figure 3.3).

Figure 3.3. *Securing expert assistance for stress management is crucial.*

Source: Cooper (2023).

This section will examine the value of professional support in stress management, outlining the different kinds of support that are available, their advantages, and the methods for obtaining them.

3.7.1. The Role of Professional Help in Stress Management

As Rickwood et al. (2007) state, seeking professional help for stress involves consulting with trained and experienced individuals who can provide guidance, support, and strategies to manage stress effectively. This approach is particularly beneficial for those experiencing severe, chronic stress or stress that significantly impacts their daily lives and overall well-being.

3.7.2. Types of Professional Help

There are various types of professional help available for stress management, each offering unique benefits (Smith, 2002):

1. **Counselling for Psychoses:** Working with a mental health expert to address emotional and psychological concerns is known as psychotherapy or talk therapy. Typical types of psychotherapy for stress reduction consist of:
    i. **Cognitive Behavioral Therapy (CBT):** It is a methodical, goal-oriented technique that assists people in recognizing and altering harmful thought patterns and stress-inducing behaviors. People can regain control over their stress reactions by adopting more constructive thought patterns and coping mechanisms.
    ii. **Mindfulness-based Stress Reduction (MBSR):** This approach helps people become more conscious of their thoughts, feelings, and bodily sensations by combining mindfulness meditation and yoga. Stress can be decreased, and emotional control improved as a result of this understanding.
    iii. **Acceptance and Commitment Therapy (ACT):** It encourages individuals to accept their thoughts and feelings rather than fighting or feeling guilty about them. It promotes commitment to personal values and actions that contribute to a meaningful life, despite stress.
2. **Psychotherapy:** Speaking with a qualified counselor who can offer assistance and direction on stress-related difficulties is the counseling process. Counselors can assist people in overcoming obstacles in life that cause stress, creating coping mechanisms, and enhancing communication skills. The duration of counseling varies according to the needs of the client.

3. **Health Care Practitioners:** Stress can occasionally cause physical symptoms or make pre-existing medical issues worse. Medical specialists, such as psychiatrists and primary care physicians, can offer thorough assessments to ascertain whether any underlying medical issues are causing stress. In order to help control the symptoms of stress, they can also prescribe drugs like anxiolytics or antidepressants.
4. **Life Guidance Experts:** Professionals known as life coaches assist clients in establishing and achieving both personal and professional objectives. By assisting clients in identifying stressors, creating action plans, and strengthening resilience, they can offer invaluable support in managing stress. Future-oriented and solution-focused thinking is common in life coaching.
5. **Assistance Teams:** Support groups provide a social setting where people can talk to others going through similar struggles about their experiences and coping mechanisms. These groups, which are led by qualified professionals, offer practical stress management guidance, emotional support, and a reduction in feelings of loneliness (Rickwood et al., 2007).

3.7.3. Advantages of Getting Professional Assistance

Seeking professional help for stress management offers numerous benefits:

1. **Knowledge and Skill:** Professionals are skilled at identifying and managing stress because of their specific education and experience. They can offer interventions and evidence-based techniques that are customized to meet the needs of each person.
2. **An Impartial Viewpoint:** Experts provide an unbiased viewpoint, devoid of prejudices and sentimental attachments that acquaintances or relatives may possess. People may be able to better understand their pressures and coping strategies thanks to this objectivity.
3. **Tailored Approaches:** Personalized evaluations and treatments are part of professional assistance. Experts assist people in identifying certain stressors and creating personalized coping mechanisms that have a higher chance of success.
4. **Support on an Emotional Level:** Professional help provides a safe and confidential space for individuals to express their feelings and concerns. This emotional support can be crucial in managing stress and preventing it from escalating into more serious mental health issues.

5. **Accountability:** Working with a professional can provide a sense of accountability, motivating individuals to implement stress management strategies and make necessary lifestyle changes.

3.7.4. How to Get Expert Assistance?

There are various processes involved in getting professional help for stress management as highlighted by Vogel et al. (2005):

1. **Determining the Requirement:** The first step is realizing when stress gets out of control and that you need to get help. Persistent feelings of overload, trouble going about everyday tasks, physical signs of stress, and an inability to control stress with self-help techniques are all indicators that you may require professional treatment.
2. **Examining Your Options:** It's crucial to look into the different kinds of expert assistance that are offered. Think about the individual requirements and preferences, such as the best way to receive therapy or counseling and whether in-person or online sessions are preferable.
3. **Locating an Experienced Expert:** Finding a qualified professional involves checking credentials, experience, and areas of expertise. Recommendations from trusted sources, such as primary care physicians, friends, or family, can be helpful. Online directories and professional associations also provide listings of qualified mental health professionals.
4. **Scheduling an Initial Consultation:** Scheduling an initial consultation allows individuals to meet with a professional, discuss their concerns, and determine if it's a good fit. This consultation can also provide an opportunity to ask questions about the professional's approach, treatment methods, and expected outcomes.
5. **Adhering to the Procedure:** Adhering to the process of obtaining professional assistance entails attending sessions regularly and actively engaging in the suggested tactics and interventions. It's crucial to have patience and an open mind because improvement could take some time (Vogel et al., 2005).

One of the most important aspects of good stress management is getting professional assistance. It offers responsibility, tailored tactics, emotional support, and access to professional information. People can choose the expert assistance that best fits their needs and tastes from the range of options offered. Recognizing the signs that professional help is needed and taking proactive steps to access it can significantly improve one's ability to manage stress and enhance overall well-being. In a world where stress is a constant companion, professional help offers a pathway to resilience and a healthier, more balanced life.

CHAPTER 4

STRESS IN THE WORKPLACE

Contents

Workplace stress is a complex phenomenon that affects worker productivity as well as that of the business. It presents a range of behavioral, psychological, and physiological symptoms that reflect the intricate relationship between a person's internal condition and the demands of their exterior profession (Glazer & Liu, 2017).

Stress at work can physiologically cause a variety of symptoms, including headaches, exhaustion, fast heartbeat, and musculoskeletal issues (Tran et al., 2020). Prolonged exposure to stressors, such as heavy workloads or deadline pressure, triggers the body's stress response, which prolongs the release of cortisol and other stress chemicals. Long-term health problems, such as immune system weakness and cardiovascular illnesses, may arise from this (Tran et al., 2020) (Figure 4.1).

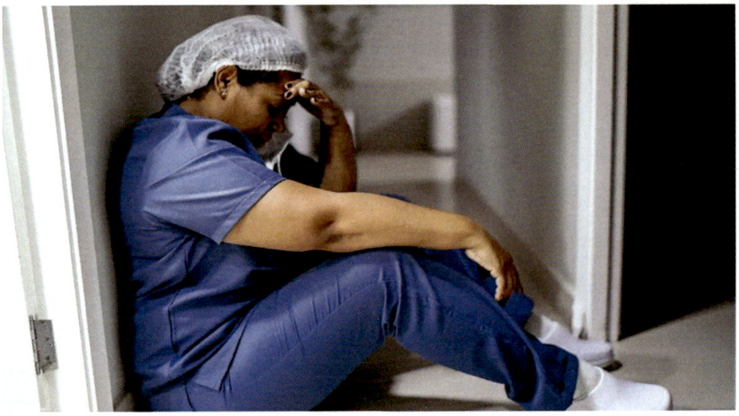

Figure 4.1. *Workforce stress in the US is common and affects career satisfaction.*

Source: Lockett (2023).

Anxiety, despair, impatience, and an overall feeling of being overwhelmed are psychological manifestations of stress (Glazer & Liu, 2017). Cognitive deficits, including issues with focus and judgment, are also prevalent. According to Tran et al. (2020), psychological effects have the potential to hinder an employee's capacity to perform well, which can lead to a decrease in both overall job performance and job satisfaction.

Workplace stress frequently causes behavioral changes in employees, including altered work habits and interpersonal relationships. Workers may show higher rates of turnover, decreased productivity, and increased absenteeism (Tran et al., 2020). Stress can also weaken relationships at work, which can result in disagreements and less teamwork. There are many different elements that contribute to workplace stress, including organizational issues like inadequate management procedures and a lack of support as well as job-specific factors like role ambiguity and workload (Glazer & Liu, 2017). A comprehensive strategy including organizational modifications, managerial development, and employee support initiatives is needed to address these problems. All things considered, workplace stress is a serious problem with wide-ranging effects on both workers and businesses. Recognizing the complexity of workplace stress and putting solutions in place to lessen its effects are essential to effective stress management.

Going forward, this chapter delves into the idea of workplace stress and offers coping mechanisms.

4.1. COMMON CAUSES OF WORKPLACE STRESS

There are several different sources of workplace stress, and each one adds to the overall stress that workers face. The academic literature has identified some of the most prevalent sources of workplace stress which are discussed below:

4.1.1. Workload and Time Pressure

Stress at work is often caused by workload and time constraints, which have a negative effect on workers' health and productivity inside the company. These pressures show up as undue demands made on workers, which frequently surpass their ability to handle them well (Hasan & Tumah, 2019).

High task volumes that demand a significant amount of time and effort result in heavy workloads, which can wear one out physically and mentally. Workers who are overworked on a regular basis lose energy and resources, which can lead to burnout (Hasan & Tumah, 2019). Chronic exhaustion affects cognitive processes like memory, attention, and decision-making, which lowers productivity and performance at work (Yaribeygi et al., 2017).

The stress brought on by a large task is increased by time constraints. Employees are forced to work at a faster pace and frequently don't have enough time to finish projects to the best of their abilities when deadlines are tight. This rushed method can result in mistakes and lower-quality work, which raises stress levels even further as workers try to fulfill deadlines while maintaining high standards (Hasan & Tumah, 2019). The ongoing pressure to fulfill deadlines breeds anxiety and a sense of urgency, which intensifies stress and leaves one feeling overburdened.

The interplay between time constraints and workload can have a detrimental impact on workers' personal lives as well. Work-life imbalance is exacerbated by long workdays and the difficulty in putting down work-related duties, which invade personal and family time (Glazer & Liu, 2017). Stress levels rise as a result of this imbalance because workers find it difficult to reconcile their home and work obligations.

Workload and time constraints can also damage relationships at work. Stressed-out workers may become agitated and less accepting of others, which can cause interpersonal problems and a bad working environment (Hasan & Tumah, 2019). These pressures' combined effects have an impact on staff members individually as well as the culture and efficacy of the company as a whole.

A multifaceted strategy is needed to address workload and time pressure, including flexible scheduling, realistic workload distribution, and resources to assist staff in managing stress. In order to foster a more positive and productive work environment, organizations need to understand how important it is to strike a balance between demands and employees' capacities (Glazer & Liu, 2017).

4.1.2. Poor Workplace Leadership and Management

Stress at work has a substantial impact on employees' mental health, job happiness, and overall performance of the organization when it is caused by poor management and leadership. Stress levels rise when workers feel mistreated, unsupported, and unclear about their responsibilities and expectations due to incompetent leadership (Hasan & Tumah, 2019).

Lack of communication is one of the main ways that bad management causes stress. Employees become apprehensive and anxious about their job security and performance when supervisors fail to effectively communicate expectations, feedback, or organizational changes (Hasan & Tumah, 2019). Because they are uncertain about their position within the company and find it difficult to accomplish ambiguous goals, employees struggle to satisfy this ambiguity and generate a stressful work environment.

Stress is also increased by a lack of managerial support. Workers look to their managers for direction, assistance, and consolation. Neglect and isolation are encouraged when managers are unapproachable, unsupportive, or uninterested in their workers' problems. Employees may experience more stress at work as a result of this lack of help since they believe they must handle obstacles and workloads on their own (Glazer & Liu, 2017).

Practices that are unjust or inconsistent are another sign of poor management. Team morale and trust are weakened when managers show partiality, apply rules inconsistently, or make arbitrary choices. These kinds of actions foster an atmosphere of unfairness, which is a powerful stressor that can result in job discontent, a decline in motivation, and a rise in plans to leave the company (Tumah & Hasan, 2019). Furthermore, workers with authoritarian leadership styles—which are marked by a high degree of control and a low degree of autonomy—are less able to manage their work processes and make decisions. Because they feel helpless and confined, workers who lack autonomy experience higher levels of stress (Tran et al., 2020).

On the other hand, research has demonstrated that participatory leadership—which gives workers a say in decisions—reduces stress by encouraging a sense of empowerment and involvement (Hasan & Tumah, 2019). Developing leadership abilities centered on cooperation, support, equity, and worker empowerment is necessary to reduce the stress brought on by incompetent management. A more encouraging and stress-free work atmosphere can be produced by training programs that improve managers' interpersonal and emotional intelligence (Tran et al., 2020).

4.1.3. Poor Interpersonal Relationships

One of the main drivers of stress in organizations is interpersonal conflict and strained relationships at work. A tense and hostile work environment is the result of these disputes, which frequently stem from incompatible personalities, divergent work styles, or rivalry for resources (Glazer & Liu, 2017).

Interpersonal problems among coworkers foster a nervous and uncomfortable environment. Constantly being in such a setting might cause emotional weariness, decreased dedication to the business, and decreased job satisfaction. Employee health and productivity may be further impacted by the physical symptoms of unresolved conflict stress, such as headaches and sleep difficulties (Glazer & Liu, 2017). Furthermore, strained relationships at work impede productive teamwork and communication. Lack of trust and cooperation hinders teamwork and can make workers feel alone and unsupported. Because social support is essential for managing work responsibilities, employees who experience social isolation feel more stressed (Tran et al., 2020).

Negative encounters with coworkers or superiors can also lower one's sense of self-worth and cause frustration and helplessness. These psychological effects raise the possibility of absence and turnover by causing a bad work experience ((Glazer & Liu, 2017). Organizational well-being can be greatly improved, and stress levels can be considerably decreased by addressing interpersonal disputes and developing healthy working relationships through team-building exercises and conflict-resolution training.

4.1.4. Lack of Autonomy

A major cause of stress at work is a lack of control and autonomy, which compromises workers' psychological health and job happiness. Employees feel powerless and frustrated when they have little control over their work assignments, timetables, or decision-making procedures. Increased stress and anxiety may result from this decreased autonomy (Tran et al., 2020).

Decision flexibility, which includes both skill discretion and decision power, is a crucial component of work control. Workers who have little discretionary control frequently find their work boring and uninteresting, which causes them to become emotionally and cognitively disengaged (Tran et al., 2020). Stress levels are further increased when there is a sense of unpredictability and instability in one's work environment and activities.

Since employees can customize their work procedures to better suit their preferences and strengths, those with greater autonomy are likely to have lower stress levels and increased job satisfaction (Tran et al., 2020). On the other hand, a lack of control can increase the stress on the body and mind, leading to negative health consequences like despair, anxiety, and burnout (Glazer & Liu, 2017).

Organizations should foster a work environment that appreciates employee input and offers chances for increased autonomy and flexibility in order to lessen these stressors. Giving workers greater freedom creates a more interesting and encouraging work atmosphere, which improves productivity and general well-being (Tran et al., 2020).

4.2. EFFECTS OF WORKPLACE STRESS ON HEALTH

Workplace stress can have numerous adverse effects on health, impacting both physical and mental well-being. Significant health effects of workplace stress, as documented in scholarly research, are:

4.2.1. Cardiovascular Diseases

Workplace stress affects employees' health through a variety of intricate physiological and behavioral processes, making it a significant risk factor for the development of cardiovascular problems. Cardiovascular diseases (CVDs) can result from a series of physiological reactions brought on by long-term exposure to work-related stresses such as excessive job expectations, poor control, and insufficient support (Glazer & Liu, 2017).

The stimulation of the sympathetic nervous system and the hypothalamic-pituitary-adrenal (HPA) axis is one of the main ways that stress affects cardiovascular health. According to Kivimäki & Steptoe (2018), prolonged stress causes the body to release stress hormones including cortisol and adrenaline, which raise blood pressure and heart rate. Persistently experiencing this elevated level of arousal can lead to hypertension disorders, which are a major risk factor for stroke and heart disease.

Furthermore, actions brought on by stress raise the risk of cardiovascular disease. High-stress levels at work increase the likelihood of harmful habits like smoking, eating poorly, not exercising, and drinking too much alcohol (Kivimäki & Steptoe, 2018). These habits increase the chance of developing diseases like atherosclerosis, which is characterized by fatty deposits clogging the arteries and raising the risk of heart attacks and other CVDs.

Moreover, endothelial function, which is essential for vascular health, might be hampered by stress. Atherosclerosis and other cardiovascular problems are preceded by endothelial dysfunction, which is typified by decreased vasodilation and elevated inflammatory responses (Turner et al., 2020). Stress-induced chronic inflammation increases the risk of coronary heart disease by facilitating the accumulation of plaques in the arteries.

Workplace psychosocial stressors like job uncertainty and conflicts also contribute to the development of CVDs. According to Kivimäki & Steptoe (2018), these stressors have the potential to cause chronic anxiety and sadness, both of which are linked to an elevated risk of cardiovascular disease.

Harmful habits, physiological changes, and psychological variables all contribute to the significant impact that job stress has on cardiovascular health. To reduce these risks and encourage employees' cardiovascular health, it is essential to have supportive work environments and effective stress management techniques.

4.2.2. Immunosuppression

Stress in the workplace weakens the immune system, reducing the body's capacity to fight off infections and making people more prone to disease. Employee health outcomes are impacted by this intricate relationship between stress and the immune system, which involves both physiological and psychological factors (Dhabhar, 2014).

Prolonged work-related stress triggers the sympathetic-adrenal-medullary (SAM) system and the HPA axis, which prolongs the release of catecholamines like noradrenaline and adrenaline as well as glucocorticoids like cortisol (Dhabhar, 2014). While high cortisol levels are initially helpful during acute stress reactions, they eventually become harmful if they persist. By preventing the synthesis and function of cytokines, which are essential for coordinating the immunological response, cortisol has immunosuppressive effects (Dhabhar, 2014).

Moreover, long-term job stress may cause an imbalance in immune cell function. Stress is specifically linked to decreased Th1 T-helper cell activity—

which is crucial for cell-mediated immunity—and increased Th2 T-helper cell activity, which is more focused on antibody-mediated immunity. According to Liu et al. (2017), this imbalance may lead to a diminished ability to fight off infections and an increased risk of allergic reactions.

Through behavioral pathways, psychological stressors at work, such as job uncertainty, excessive demands, and lack of support, also worsen immune system suppression. Chronically stressed workers are more prone to engage in harmful habits including smoking, eating poorly, getting too little sleep, and not exercising—all of which worsen immune function (Dhabhar, 2014).

Studies show that stress at work can also delay the healing of wounds and decrease the efficiency of immunizations, demonstrating the real-world effects of immune suppression brought on by stress on health outcomes (Barsotti et al., 2020). These negative consequences underline how important it is to implement interventions including organizational modifications, stress management courses, and creating a supportive work environment in order to lessen workplace stress.

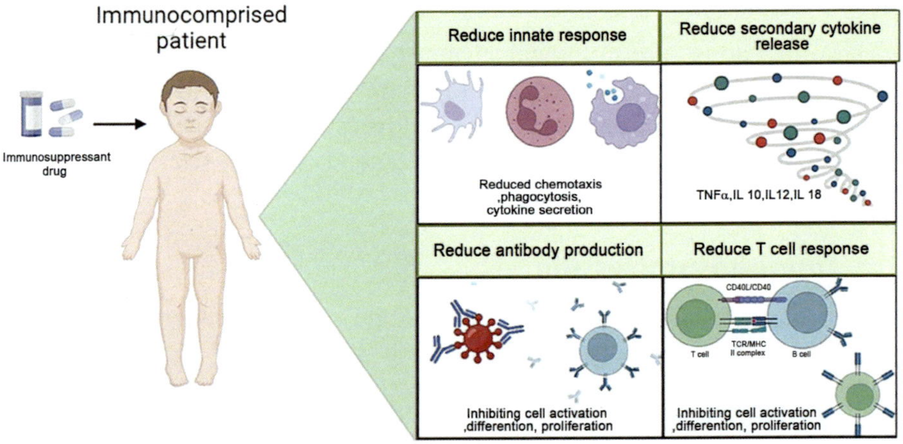

Figure 4.2. *Adverse effects of immunosuppression.*

Source: Sadeghi et al. (2023).

Psychological stressors at work, such as employment insecurity, excessive demands, and lack of support, also exacerbate immune system suppression through behavioral pathways. Workers who experience ongoing stress are more likely to adopt unhealthy behaviors, such as smoking, eating badly, sleeping too little, and not exercising, all of which impair immune function (Dhabhar, 2014).

Research indicates that occupational stress can also impede wound healing and reduce vaccine effectiveness, illustrating the practical implications of immune suppression due to stress on health outcomes (Barsotti et al., 2020). These unfavorable outcomes highlight how crucial it is to put interventions—such as organizational changes, stress management training, and fostering a supportive work environment—into place in order to reduce workplace stress (Figure 4.2).

4.2.3. Mental Health Disorders

Stress at work has a significant impact on employees' psychological health and general functioning, which is why it is a significant contributor to mental health illnesses. The likelihood of acquiring illnesses like anxiety, depression, and burnout is greatly increased by prolonged exposure to stressors including heavy workloads, deadline pressure, and a lack of support (McEwen, 2017).

There is a connection between mental health issues and working stress that is mediated by both psychological and physiological systems. According to physiological principles, long-term stress triggers the HPA axis, which causes cortisol and other stress hormones to be secreted continuously (McEwen, 2017). Long-term cortisol exposure has an impact on the hippocampus and prefrontal cortex, two brain regions involved in mood regulation. These changes may result in structural and functional alterations that make people more susceptible to anxiety and depressive disorders (McCraty & Atkinson, 2012).

Psychologically, occupational stress increases anxiety and a state of perpetual vigilance, which makes it harder for workers to handle day-to-day difficulties. Resilience and self-efficacy are crucial components of mental health maintenance that can be undermined by ongoing stress (McEwen, 2017). Core symptoms of depression, such as despair and hopelessness, can arise when workers believe their work environment is unmanageable and overpowering (Turner et al., 2020).

In addition, the social aspect of the job is quite important. Stress is made worse by the absence of social support from coworkers and superiors, which raises the risk of mental health problems. Emotional distress can be exacerbated by interpersonal disputes and isolation, which heighten feelings of alienation and loneliness (McEwen, 2017).

Employees under stress may act out by abusing substances, eating poorly, or getting too little sleep, which exacerbates mental health issues (McCraty & Atkinson, 2012). These actions set up a vicious cycle in which unhealthy habits brought on by stress worsen psychological discomfort, which in turn worsens mental health consequences.

In summary, a mix of physiological alterations, psychological suffering, and social isolation is how occupational stress results in mental health illnesses. Creating a supportive work environment, putting stress management programs into place, and giving employees access to mental health resources are all effective ways to reduce workplace stress. These tactics are crucial for improving workers' mental health and creating a happier, more productive work environment.

4.2.4. Sleep Disturbance

Stress at work has a great impact on sleep disorders, which include insomnia, poor sleep quality, and irregular sleep patterns. Scholarly studies emphasize how closely work-related pressures and their effects on sleep are related. A study by Turner et al. (2020) found that physiological reactions that disrupt the onset and maintenance of sleep can be triggered by stressors such as work-related demands, interpersonal disputes, and workload pressures.

The disruption of the HPA axis is one way by which sleep problems and occupational stress are linked. Chronic stress causes the circadian rhythms necessary for sleep-wake cycles to be disrupted by the release of cortisol (Turner et al., 2020). Additionally, the psychological strain brought on by work-related stressors frequently results in concern and rumination, which lengthens sleep latency and decreases sleep efficiency (McCraty & Atkinson, 2012).

Moreover, a continuous state of stress can start a vicious cycle in which getting too little sleep makes you feel even more stressed, which is a negative feedback loop. Over time, this cycle impairs not only mood regulation and cognitive abilities but also physical health (Turner et al., 2020).

In order to lessen the negative impacts of workplace stress on worker productivity and well-being, it is essential to comprehend how it causes sleep disorders. Organizational leaders and employers should place a high priority on measures that lessen workplace stress and foster a welcoming workplace that supports restful sleep patterns.

4.3. STRATEGIES FOR MANAGING WORKPLACE STRESS

Managing workplace stress is crucial for maintaining employee well-being and organizational productivity. Here are some strategies supported by scholarly sources that can be used by both employers and employees in managing workplace stress:

4.3.1. Promoting Work-Life Balance

Several academic research have demonstrated the importance of work-life balance in reducing workplace stress. Work-life balance is the state of equilibrium that people pursue between their obligations in their personal and professional lives. It includes a variety of elements, including time management, flexible work schedules, and psychological disengagement from work during non-working hours (Glazer & Liu, 2017).

Firstly, encouraging work-life balance lowers the risk of employee burnout, which in turn lessens stress. Long-term work stress frequently results in burnout, which is defined by emotional depletion, depersonalization, and decreased personal accomplishment (Foy et al., 2019). Organizations can enhance overall well-being and job satisfaction by preventing the exhaustion of physical and emotional resources by providing employees with sufficient time for personal activities and relaxation.

Second, encouraging work-life balance leads to better results for mental health. According to research, long-term work stress can cause depression, anxiety, and other mental health problems (Foy et al., 2019). Giving workers the chance to engage in extracurricular activities and time management skills enhances psychological resilience and lessens the damaging effects of work-related stress.

Initiatives for work-life balance can also improve retention and productivity, two important organizational objectives. Workers are more likely to be engaged and dedicated to their work if they believe that their company supports their personal well-being (Glazer & Liu, 2017). The significance of establishing a work environment that prioritizes and supports employees' holistic needs is highlighted by this positive association.

Encouraging work-life balance is a deliberate approach to stress management and improving overall organizational performance, not only a question of personal preference on the part of employees. Organizations may foster a more robust and efficient workforce by acknowledging the interdependence of work and personal life and enacting policies that promote it.

4.3.2. Training and Provision of Resources

Scholarly research supports that a major factor in helping people manage their stress at work is providing them with appropriate resources and training. According to Foy et al. (2019), suitable resources include both material and intangible supports required for job performance, whereas training refers to the acquisition of information, skills, and abilities relevant to job activities.

First of all, training gives workers the abilities and information they need to do their jobs well, which helps lessen stress brought on by insecurity and ineptitude. Studies indicate that workers who obtain thorough training see themselves as more capable and self-assured in their positions, which reduces stress associated with work (Foy et al., 2019). Training courses that emphasize time management, resilience building, and stress management strategies can equip workers to handle the rigors and demands of the workplace. Second, the key to lowering stress is to have an adequate supply of tools, technology, and support systems. Inadequate resources or outdated equipment may have a detrimental influence on employee unhappiness and productivity (Hasan & Tumah, 2019). Companies that invest in modern technology, ergonomic workspaces, and auxiliary infrastructure create an environment where employees can do their duties more effectively and with less stress from operational constraints.

Furthermore, in order to adjust to changing work settings and technological breakthroughs, it is imperative to provide continuous training and update resources. Opportunities for continuous learning not only improve job skills but also help employees feel committed to the organization and grow personally (Foy et al., 2019). In a similar vein, frequent resource updates guarantee that staff members have access to the newest instruments and support networks, enhancing their productivity and reducing stress brought on by antiquated procedures.

Incorporating thorough training programs and making sure there are enough resources available are essential tactics for handling stress at work. Through the provision of employee development opportunities and the maintenance of encouraging work environments, firms can improve worker happiness, output, and general health.

4.3.3. Encouraging Social Support

Strong scholarly research supports the important strategy of promoting social support among employees as a means of controlling stress. Social support is the umbrella term for the tools that people offer to mitigate the negative impacts of stressors. These tools might include emotional, informational, and practical help (Foy et al., 2019).

To begin with, social support helps people feel less alone and more like they belong, which protects them from stress. Empirical studies suggest that workers who perceive robust social support networks inside their workplace are more capable of managing job demands and resolving interpersonal problems (Foy et al., 2019). When coworkers and managers show compassion, understanding, and support, they foster a positive environment where people feel appreciated and validated, which lowers stress levels.

Second, social support strengthens one's capacity for problem-solving and resilience in the face of difficulties at work. Employees can learn new viewpoints and stress-reduction techniques by working together on projects, sharing experiences, and exchanging knowledge (Turner et al., 2020). Employees are more likely to use effective stress management techniques when they operate in cohesive teams and cooperative work cultures that encourage constructive criticism and support from one another. Organizational initiatives that encourage mentorship programs, open communication lines, and team-building exercises all help to foster a supportive and friendly culture. According to Foy et al. (2019), these kinds of programs foster a feeling of community in the workplace in addition to strengthening interpersonal ties. Socially connected workers are more inclined to ask for and offer assistance, creating a positive feedback loop that improves general well-being and job satisfaction.

Therefore, creating social support networks at work is critical to stress management and the development of a positive company culture. Through the adoption of supportive practices and an appreciation of the value of interpersonal relationships, organizations can foster employee engagement, build resilience, and lessen the detrimental effects of stress on both individual and organizational outcomes.

4.3.4. Stress Reduction Programs

According to academic studies, stress management programs are essential for treating and reducing workplace stress. These programs include a range of treatments intended to provide staff members with the knowledge, abilities, and tools they need to manage and lower their stress levels (Sharma & Rush, 2014).

First of all, education and training sessions that increase knowledge of the origins and effects of stress are commonly included in stress management programs. These programs enable people to identify early indicators of stress and take proactive coping mechanisms by improving staff members' comprehension of stressors and their physiological and psychological effects (Sharma & Rush, 2014). With this information, staff members may better handle stress and lessen its negative impacts on their health and the effectiveness of the company.

Second, to encourage stress reduction and improve resilience, these programs frequently include mindfulness exercises, relaxation techniques, and cognitive-behavioral approaches (Khoury et al., 2015). It has been demonstrated that methods including deep breathing exercises, meditation, and time management can reduce the symptoms of stress and enhance mental health in general (Sharma & Rush, 2014).

Programs for stress management also show a company's dedication to its employees' well-being, which helps to foster a positive workplace culture. Businesses that fund these initiatives let employees know that stress is recognized and that there are tools available to help them manage it well (Sharma & Rush, 2014). This proactive strategy improves employee morale and job satisfaction in addition to creating a positive work environment.

To sum up, stress management initiatives are crucial resources for businesses looking to support worker well-being and output. Through the provision of knowledge, skill development, and supportive resources, these programs facilitate effective stress management, resilience building, and increased organizational efficacy as a whole.

4.4. CREATING A HEALTHY WORK ENVIRONMENT

Stress has become an ever-present concern in today's fast-paced and constantly evolving workplace. Prolonged stress can affect one's emotional and professional well-being by causing several health problems, such as anxiety, depression, and cardiovascular illnesses. Thus, stress reduction and the development of a happy, productive staff depend on having a healthy work environment. This paper examines the essential tactics for establishing this kind of setting, emphasizing organizational, psychological, and physical elements (Brunges & Foley-Brinza, 2014).

4.4.1. Workspace Design and Ergonomics in the Physical Environment

Physical strain and discomfort are greatly reduced by an ergonomic workspace, which can eventually result in stress. Employers are responsible for making sure workstations are built to encourage proper posture and lessen injuries from repetitive strain. Sit-stand workstations, movable chairs, and strategically positioned monitors and keyboards can all significantly improve hence making a significant difference (Berndt & Ortelli, 2023).

1. **Air Quality and Lighting:** For a workplace to be healthy, proper lighting and air quality are essential. Enough ventilation keeps indoor contaminants from building up, while natural light boosts mood and vitality. Employers who want to guarantee a cozy and healthful work environment should invest in HVAC systems and high-quality lighting solutions.
2. **Control of Noise:** One of the biggest causes of stress might be excessive noise. Employers can reduce noise pollution by establishing quiet zones, soundproofing, or using white noise generators. Permitting workers to use noise-canceling headphones can also improve concentration and lower stress levels (Berndt & Ortelli, 2023).

4.4.2. Psychological Environment

1. **Support for Mental Health:** Reducing job stress requires offering mental health care. Access to counseling services, whether provided

internally or by outside providers, should be provided by employers. Wellness initiatives, stress management classes, and mental health days can all be beneficial resources for staff members.

2. **Promoting Work-Life Harmony:** The lines separating work and personal life are respected in a healthy work environment. In order to help employees maintain a work-life balance and prevent stress and burnout, employers should offer flexible working hours, remote work choices, and clear policies on overtime.
3. **Acknowledgment and Benefits:** Stress can be considerably reduced, and staff morale is raised when hard work is acknowledged and rewarded. Employees can feel appreciated and driven when they receive positive reinforcement, constructive criticism, bonuses, or paid time off.

4.4.3. Organizational Environment

1. **Open and Honest Communication:** Clear and honest communication builds confidence and lessens ambiguity, which is a significant cause of stress. Transparency and respect for one another can be fostered by open-door practices, frequent updates from management, and supportive comments.
2. **Clearly Defined Roles and Duties:** Roles and responsibilities that are unclear can cause anxiety and uncertainty. Employers need to make sure that job descriptions are clear and that workers are aware of their responsibilities. Clear expectations and regular check-ins can be achieved with the support of performance reviews.
3. **Advancement of the Profession:** Possibilities for professional development can improve career possibilities and job satisfaction, which can lower stress. To support employees in advancing in their careers, employers should fund training initiatives, workshops, and mentorship programs.
4. **Team Building and Social Activities:** Strong social connections at work can provide a support system that helps alleviate stress. Organizing team-building activities, social events, and collaborative projects can strengthen workplace relationships and create a sense of community (Brunges & Foley-Brinza, 2014).

4.4.4. Stress Reduction Techniques

1. **Meditation and Mindfulness:** Stress levels can be dramatically lowered by engaging in mindfulness and meditation techniques. Employers can give staff access to a quiet area where they can

practice these skills during breaks, as well as mindfulness training and meditation programs.

2. **Exercise:** It is a well-known way to alleviate stress. Employers can promote physical activity by setting up group exercise classes, installing on-site fitness centers, or providing rewards for active commuting like showers and bike racks.
3. **Optimal Nutrition:** Stress management can be aided by eating well. Employers can encourage healthy eating by supplying wholesome snacks, giving balanced meal selections in the cafeteria, and advancing wellness initiatives that encourage a healthy diet.

4.4.5. Breaks and Downtime

Regular breaks are essential for maintaining productivity and reducing stress. Employers should encourage employees to take short breaks throughout the day and to use their vacation time fully. Creating a culture that values downtime can help prevent burnout (Brunges & Foley-Brinza, 2014).

Building a healthy workplace necessitates a multifaceted strategy that takes organizational, psychological, and physical aspects into account. Employers may create a work environment where people flourish by emphasizing ergonomics, lighting, noise management, mental health assistance, work-life balance, open communication, defined roles, professional growth, team building, and stress-reduction strategies. In addition to improving workers' health, a positive work environment raises productivity levels and fosters organizational success. It will become more crucial to keep an emphasis on fostering a positive and stress-free atmosphere as the nature of work continues to change.

4.5. BALANCING WORK AND PERSONAL LIFE

Maintaining a healthy work-life balance is among the most challenging things for individuals to do. Globalization, technology, and demanding work environments have blurred the lines between the personal and professional domains. This imbalance frequently results in elevated stress levels, which have an impact on productivity, general well-being, and mental and physical health (Williams et al., 2015). Determining how to walk this delicate balance is essential to long-term success and well-being (Figure 4.3).

Figure 4.3. *Importance of work-life balance.*

Source: Eby (2023).

4.5.1. The Modern Work Environment

The modern work environment is characterized by rapid technological advancements, constant connectivity, and a competitive global market. While these developments have brought many benefits, such as increased efficiency and flexibility, they have also introduced new stressors. Employees are often expected to be available around the clock, leading to an "always-on" culture that erodes personal time and exacerbates stress.

Although stress is a normal reaction to difficult circumstances, prolonged stress can have detrimental effects. Stress physiologically sets off the body's "fight or flight" response by releasing hormones like cortisol and adrenaline (Tran et al., 2020). Short-term benefits from this reaction notwithstanding, extended exposure to stress hormones can result in many health problems, such as weaker immune systems, heart disease, and mental health conditions including despair and anxiety.

4.5.2. The Factors That Lead to Work-Life Unbalance

The mismatch between work and personal life is caused by a number of reasons as stated by Williams et al. (2015), each of which has its own effect on stress levels:

1. **Extended Working Hours:** Long workdays don't leave much time for hobbies, family time, or leisure. Burnout, which is characterized by emotional, mental, and physical tiredness, may result from this.
2. **Lack of Control:** Stress levels might increase when there is a lack

of autonomy at work. Workers who believe they have no control over their workloads or schedules are more prone to feel anxious and unsatisfied.

3. **Unrealistic Expectations:** It can be quite stressful to live up to the high standards that you set for yourself or that your employers set for you. Tight deadlines and unrealistic goals frequently result in high levels of stress and low productivity.
4. **Workplace Culture:** A culture that glorifies overworking and undervalues personal time can make employees feel guilty for taking breaks or setting boundaries. This culture perpetuates stress and diminishes overall well-being.
5. **Technological Intrusion:** Although technology allows for flexibility and remote work, it also makes it more difficult to distinguish between work and personal life. It can be challenging to step back and rejuvenate when emails, messages, and notifications are coming in nonstop (Williams et al., 2015).

4.5.3. Methods for Juggling Your Personal and Professional Lives

It takes deliberate effort and smart preparation to strike a balance between work and personal life. The following are a few methods to reduce stress and enhance this balance:

1. **Set Boundaries:** Clearly define the lines separating work and personal time. Set aside specified times for work and make sure you keep to them. Share these limits with coworkers and managers to set expectations and minimize disruptions during personal time.
2. **Prioritize Self-Care:** Self-care is essential for managing stress. Regular exercise, healthy eating, adequate sleep, and engaging in hobbies or leisure activities can significantly improve physical and mental well-being. Schedule self-care activities just as you would work meetings to ensure they are not neglected.
3. **Time Management:** Effective time management can alleviate stress and improve productivity. Use tools like calendars, to-do lists, and time-blocking techniques to organize tasks and allocate time efficiently. Prioritize tasks based on their importance and deadlines, and avoid multitasking, which can reduce efficiency and increase stress.
4. **Learn to Say No:** Overcommitting is a common source of stress. Learn to say no to additional tasks or responsibilities that would

overload your schedule. Be realistic about your capacity and prioritize tasks that align with your goals and values.

5. **Leverage Technology Wisely:** Use technology to enhance productivity and flexibility without letting it intrude on personal time. Set boundaries for checking emails and messages outside of work hours. Utilize productivity apps to streamline tasks and reduce the time spent on administrative work.
6. **Seek Support:** Effective stress management requires a robust support network. Talk to your loved ones, coworkers, or friends about your struggles so they can provide support, motivation, or guidance. To create efficient stress management techniques, think about obtaining expert assistance, such as coaching or counseling.
7. **Engage in Mindfulness Practices:** By encouraging relaxation and mental clarity, mindfulness practices like yoga, meditation, and deep breathing can help manage stress. Include mindfulness exercises in your routine to help you focus better and feel less anxious.
8. **Take Regular Pauses:** Taking regular pauses while working will help you stay productive and avoid burnout. Taking quick breaks to stretch, go for a walk, or do anything unrelated to work will help you mentally and physically recharge and increase your productivity.
9. **Flexible Work Arrangements:** If possible, negotiate flexible work arrangements with your employer. Options like remote work, flexible hours, or compressed workweeks can provide a better balance between work and personal responsibilities, reducing stress.
10. **Evaluate and Modify:** Consistently evaluate your work-life equilibrium and pinpoint opportunities for enhancement. Consider what is and is not working and be prepared to make changes. The task of juggling business and personal obligations calls for constant effort and flexibility (Williams et al., 2015).

4.5.4. The Employers' Role

To support employees' work-life balance and lower stress levels, employers are crucial. Employers can improve worker satisfaction, productivity, and well-being by encouraging a healthy work environment and work-life policies (Tran et al., 2020).

1. **Encourage a Positive Culture:** Establish a work environment where employees' well-being is valued. Promote establishing boundaries, acknowledging accomplishments, and taking breaks. Stress the value of personal time and discourage working excessive overtime.

2. **Offer Flexibility:** Give employees flexible work schedules that meet their various needs. Flexible employees can handle their duties better and feel less stressed.
3. **Provide Resources:** Make available resources like stress management courses, wellness initiatives, and employee assistance programs (EAPs). Giving workers access to professional development opportunities and mental health support can help them learn appropriate stress management techniques.
4. **Promote Open Communication:** Create an atmosphere where staff members feel free to talk about difficulties and ask for help. Early problem detection and cooperative problem-solving can be facilitated by open communication.
5. **Set a Good Example:** Executives should exhibit a positive work-life balance. Setting boundaries and putting their well-being first creates a good example for staff members to follow.

Balancing work and personal life in the modern world is a complex challenge, but it is achievable with intentional strategies and support.

4.6. SEEKING SUPPORT IN THE WORKPLACE

Workers deal with a variety of difficulties, including personal problems, interpersonal conflicts, and high workloads and deadlines. Overcoming these obstacles can have a big impact on someone's productivity, job happiness, and mental health. Seeking assistance from coworkers is essential to preserving a good work-life balance and creating a supportive and productive work atmosphere (Lundqvist et al., 2018). The significance of seeking assistance, the kinds of help that are accessible, and useful methods for gaining and making use of this support will all be covered.

4.6.1. The Importance of Seeking Support

4.6.1.1. Mental Wellness

Maintaining mental health and well-being is one of the most important reasons to ask for help at work. If work-related stress is not properly treated, it can result in burnout, anxiety, and depression. Employees who receive support from coworkers, managers, and mental health specialists may find much-needed respite and learn effective coping skills (Lundqvist et al., 2018).

4.6.1.2. Workplace Contentment and Effectiveness

Workers are more likely to be happier in their jobs when they feel supported. Since motivated, engaged, and devoted workers are more likely to feel supported at work, this sense of fulfillment can lead to improved job performance. Furthermore, a positive workplace culture can lower attrition and raise employee retention.

4.6.1.3. Building a Positive Work Culture

A workplace where support is readily available fosters a positive work culture. Employees feel valued and respected, which can lead to increased collaboration, creativity, and innovation. A positive work culture also attracts top talent, making it easier for organizations to recruit and retain skilled employees (Hubbard et al., 2018).

4.6.2. Types of Support in the Workplace

1. **Support on an Emotional Level:** Colleagues who receive emotional support from their peers benefit from encouragement, empathy, and understanding. Peers, managers, and workplace counselors are some sources of this kind of support. It aids with stress management, resilience building, and a stronger sense of community inside the workplace for staff members.
2. **Informational Assistance:** Informational support is the exchange of expertise, counsel, and direction to assist colleagues in problem-solving or decision-making. This assistance, which is essential for skill development and job advancement, might come from mentors, supervisors, or professional networks.
3. **Essential Assistance:** Tangible aid, such as task assistance, resource provision, or flexible work schedules, is referred to as instrumental support. Employees who receive this kind of support may balance their personal and professional obligations and feel less stressed about their burdens.
4. **Appraisal Support:** It involves providing constructive feedback, recognition, and validation. This support helps employees understand their strengths and areas for improvement, boosting their confidence and motivation (Lundqvist et al., 2018).

4.6.3. Techniques for Requesting Assistance

1. **Acknowledge Your Need for Assistance:** Acknowledging when support is required is the first step towards obtaining it. Workers need

to be aware of any indications of stress, burnout, or deteriorating productivity. Early detection of these symptoms can help to avert more serious problems and enable prompt care.

2. **Determine Which Resources are Available:** Workers ought to become acquainted with the assistance resources offered by their company. Employee assistance programs (EAPs), mental health services, mentorship programs, and chances for professional growth are a few examples of these resources. When needed, employees can seek the necessary support by being aware of what is available.
3. **Communicate Openly and Honestly:** Effective communication is crucial when seeking support. Employees should feel comfortable discussing their needs and concerns with their supervisors or colleagues. Open and honest communication fosters trust and ensures that the support provided is tailored to the individual's needs.
4. **Make Use of Peer Assistance:** The benefits of peer support are numerous since peers are usually aware of the difficulties and demands that come with the job. Workers should ask dependable colleagues for guidance, support, and cooperation. A more welcoming and encouraging work atmosphere can be achieved by developing a strong network of support among coworkers.
5. **Make Use of Expert Assistance:** Help from a professional may be required in some situations. Workers should not be afraid to ask for help from experts such as career coaches or mental health specialists. Expert assistance can offer specific direction and assistance that might not be found in the workplace.
6. **Set Boundaries:** Setting boundaries is an essential aspect of seeking support. Employees should communicate their limits and ensure that they are not taking on more than they can handle. Setting boundaries helps prevent burnout and ensures that employees can maintain a healthy work-life balance.
7. **Practice Self-Care:** Self-care is a vital component of seeking support. Employees should prioritize activities that promote their well-being, such as exercise, hobbies, and spending time with loved ones. Practicing self-care helps employees recharge and better manage workplace stress (Hubbard et al., 2018).

4.6.4. Promoting a Helpful Work Environment

Some of the ways of promoting a helpful work environment for employees and their managers are through the following methods (Thomas et al., 2022).

1. **Dedicated Leadership:** The development of a positive work atmosphere is mostly the responsibility of leaders. They ought to place a high priority on the welfare of their staff members and show that they are willing to help. By encouraging work-life balance, having frank conversations about mental health, and emphasizing the value of getting treatment, leaders can set an example.
2. **Education and Training:** Employers should fund training and development initiatives that give staff members the ability to both offer and receive assistance. These courses may cover conflict resolution, effective communication, and mental health awareness.
3. **Establishing Secure Areas:** It's critical to establish safe spaces where staff members can voice their worries. Employers can create discreet counseling services, support groups, and frequent check-ins to make sure staff members feel comfortable asking for assistance.
4. **Promoting Work-Life Balance:** Promoting work-life balance is a key aspect of a supportive work environment. Organizations can implement policies such as flexible work hours, remote work options, and adequate vacation time to help employees manage their professional and personal responsibilities.
5. **Recognizing and Rewarding Supportive Behavior:** Recognizing and rewarding employees who demonstrate supportive behavior can reinforce the importance of support in the workplace. Organizations can acknowledge these employees through awards, public recognition, or other incentives (Thomas et al., 2022).

4.7. RECOGNIZING SIGNS OF BURNOUT IN THE WORKPLACE

Burnout has become an increasingly prevalent issue in several workplaces. Characterized by chronic physical and emotional exhaustion, cynicism, and feelings of reduced professional efficacy, burnout can have severe implications not only for the affected individuals but also for their organizations (Bridgeman et al., 2018). Recognizing the signs of burnout early is crucial for mitigating its effects and promoting a healthy, productive workplace.

4.7.1. What Is a Burnout?

There is more to burnout than just sporadic tension or exhaustion. Burnout is defined by the World Health Organization (WHO) as an occupational phenomenon brought on by ongoing workplace stress that has not been effectively managed. Usually, burnout appears in three different ways:

i. **Exhaustion:** A sense of being emotionally and physically spent.

ii. **Cynicism:** Taking on a pessimistic or disengaged outlook towards one's work.

iii. **Inefficacy:** Feeling less capable and accomplished than before (Bridgeman et al., 2018).

4.7.2. Understanding the Symptoms of Burnout

1. **Tiredness, Both Physical and Emotional:** Persistent tiredness is one of the most obvious indicators of burnout. Workers may complain about being exhausted even after getting enough sleep and finding it difficult to muster the energy to finish their jobs. As the immune system deteriorates due to prolonged exhaustion, workers may become more prone to colds, the flu, and other infections.
2. **Diminished Efficiency and Output:** Job performance frequently shows a discernible drop as a result of burnout. Workers could find it challenging to focus, decide what to do, or finish work quickly. Physical weariness and cognitive impairment may both contribute to this decrease in productivity, as burnout's mental toll impairs one's capacity for concentration and clear thought.
3. **A Rise in Cynicism and Disengagement:** A change in perspective about work is one of the main signs of burnout. Workers may grow more and more pessimistic, seeing their jobs and coworkers negatively. This disengagement might show itself as a general feeling of disengagement from their obligations, as well as a loss of excitement for initiatives and decreased interest in teamwork.
4. **Modifications in Behavior and Emotion:** Significant alterations in an employee's emotional state and behavior may result from burnout. People might become more easily agitated, frustrated, and prone to mood fluctuations. In addition, they could avoid social situations at work and instead choose to spend their time alone rather than interacting with coworkers or close friends.
5. **Physical Symptoms:** The stress associated with burnout can cause various physical symptoms, including headaches, stomach issues, and muscle tension. These symptoms often go unnoticed as being related to burnout, as they are commonly attributed to other health conditions.
6. **Decreased Job Satisfaction:** Employees experiencing burnout often report a significant decrease in job satisfaction. The sense of purpose and fulfillment they once found in their work diminishes, leading to a lack of motivation and a sense of helplessness. This

decrease in satisfaction can further exacerbate feelings of inefficacy and contribute to a downward spiral of declining mental health (Tran et al., 2020) (Figure 4.4).

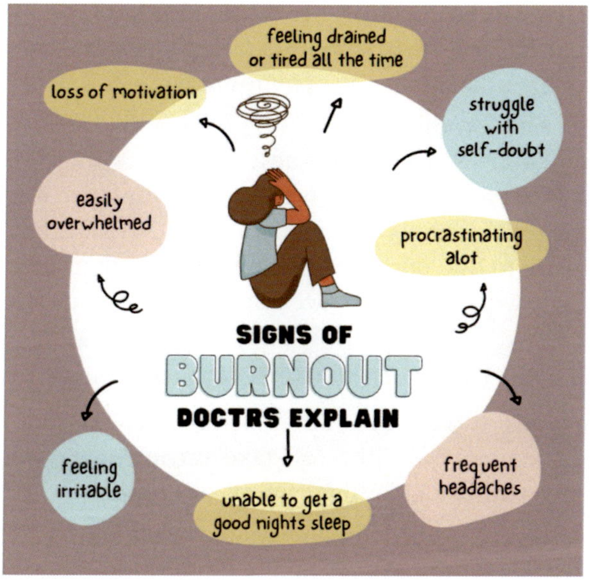

Figure 4.4. *Common symptoms of burnout.*

Source: Masson (2023).

4.7.3. Reasons for Fatigue

As reported by Bridgeman et al. (2018), recognizing the symptoms of burnout and putting effective solutions in place requires an understanding of its origins. Burnout at work is caused by several things, which include:

1. **Tasks and Hours of Work:** Long hours and an excessive workload are the main causes of burnout. Employees who are routinely overworked and overextended find it difficult to relax and recuperate, which can result in long-term stress and fatigue.
2. **Insufficient Authority and Freedom:** Burnout is more likely to occur in workers who believe they have little control over their work or limited autonomy in making decisions. Helplessness and frustration can result from not being able to affect job processes or outcomes.
3. **Insufficient Rewards and Recognition:** A lack of recognition and rewards for hard work can significantly impact employee morale.

When employees feel that their efforts go unnoticed or unappreciated, their motivation and job satisfaction can decline, increasing the risk of burnout.

4. **Negative Relationships at Work:** Burnout can be exacerbated by conflict, a lack of support, and poor communication at work. Retaining a healthy work environment and reducing stress levels require strong, supportive connections with bosses and coworkers.
5. **Incongruous Skill and Values:** Burnout is more likely to happen when there is a mismatch between an employee's values, abilities, and job requirements (Bridgeman et al., 2018). Employees may find it difficult to find purpose and fulfillment in their jobs if they believe that their work does not match their objectives or strengths.

4.7.4. Preventing and Addressing Burnout

Recognizing the signs of burnout is the first step in addressing and preventing it. Employers and employees alike can take proactive measures to create a healthier work environment and reduce the risk of burnout.

1. **Encourage Equilibrium between Work and Life:** Promoting a healthy work-life balance among employees is essential to averting burnout. Policies that encourage remote work choices, flexible work schedules, and adequate vacation time for rest and recovery can be put in place by employers.
2. **Offer Resources and Assistance:** In order to manage burnout, access to mental health resources and support networks is crucial. To help workers manage stress and preserve their well-being, employers might provide employee assistance programs (EAPs), counseling services, and stress management courses.
3. **Promote a Healthy Workplace:** Reduced risk of burnout can be achieved by fostering a good, inclusive work environment where employees feel appreciated and supported. Reward systems, open lines of communication, and regular team-building exercises can all contribute to the development of a positive, encouraging work environment.
4. **Promote Autonomy and Control:** Giving workers more freedom and authority over their work can improve job satisfaction and lower burnout. Employers can provide opportunities for skill development, permit flexible work schedules, and involve workers in decision-making processes.
5. **Keep an Eye on Expectations and Workloads:** Preventing burnout requires routine workload assessment and management. Employers

must make sure that duties and expectations are reasonable and that workers have the tools and assistance they need to do their jobs well.

6. **Deal with and Settle Disputes:** Resolving disputes and promoting positive working relationships are critical to reducing stress and averting burnout. It is recommended that employers foster a culture of respect and collaboration, offer conflict resolution training, and support transparent communication.

7. **Align Roles with Strengths and Values:** Ensuring that employees' roles align with their skills, strengths, and values can enhance job satisfaction and reduce burnout. Employers should strive to match employees with tasks and responsibilities that align with their personal and professional goals (Smith, 2002).

To sum up, stress management initiatives are crucial resources for businesses looking to support worker well-being and output. Through the provision of knowledge, skill development, and supportive resources, these programs facilitate effective stress management, resilience building, and increased organizational efficacy as a whole.

CHAPTER 5

STRESS AND RELATIONSHIPS

Contents

Stress and romantic relationships are closely related and have a variety of mutually reinforcing effects. Stress, which is characterized as a condition of mental or emotional strain brought on by difficult or demanding situations, has a big influence on how well love relationships work. Stress can lower relationship satisfaction by lowering emotional availability, raising conflict, and decreasing communication, claim Kiecolt-Glaser & Wilson (2017). According to research, negative interaction patterns can be made worse by chronic stress in particular, which can cause relational dynamics to deteriorate cyclically (Randall & Bodenmann, 2017) (Figure 5.1).

Figure 5.1. *Stress affects your relationships and health.*

Source: Formen Health (2024).

Relationship strain can also be increased by the stress contagion effect, in which the stress of one spouse influences the other. This phenomenon highlights the value of dyadic coping, in which spouses help one another manage stress, improving relationship satisfaction and resilience (Kiecolt-Glaser & Wilson, 2017). On the other hand, a stable and nurturing romantic relationship can act as a stress reliever by offering emotional support, boosting emotions of security, and encouraging flexible coping mechanisms (Randall & Bodenmann, 2017).

The association between romantic relationships and stress underscores the need for stress management to be addressed in couple therapy and other relationship-healthy therapies. Comprehending these processes is essential to formulating efficacious tactics to alleviate the deleterious consequences of stress on love relationships.

Going forward, this chapter will explore the role of stress in romantic relationships and discuss effective management strategies to promote relationship health.

5.1. IMPACTS OF STRESS ON RELATIONSHIPS

Chronic stress negatively affects partner relationships in several ways which are discussed in subsections.

5.1.1. Impaired Communication

Stress has a substantial negative impact on communication in relationships through a number of harmful ways. Cognitive overload is a common phenomenon in stressed individuals, impeding their capacity to effectively process information and respond appropriately (Kiecolt-Glaser & Wilson, 2017). Stressed people often mistake their partner's words or intentions, which can cause unnecessary conflict and strain in their relationships. This cognitive pressure can also result in misunderstandings (Simpson & Rholes, 2017).

Furthermore, stress causes physiological reactions that might amplify emotional reactivity, such as an accelerated heart rate and cortisol levels (Kiecolt-Glaser & Wilson, 2017). This increased sensitivity frequently shows up as defensiveness or impatience, which hinders productive communication and intensifies confrontations. According to Randall & Bodenmann (2017), people who are under stress are more prone to use negative communication styles including stonewalling, criticism, and contempt. These behaviors are harmful to the stability and happiness of relationships.

According to Kiecolt-Glaser & Wilson (2017), stress also lowers emotional availability, which makes it difficult for people to successfully offer or receive emotional assistance. This emotional detachment can exacerbate relationship problems by fostering a feeling of loneliness and alienation. According to Simpson & Rholes (2017), treatments that are meant to improve relational outcomes must include both effective stress management and training in communication skills. It might be helpful to establish techniques to improve relational resilience and encourage healthier interactions by having a better understanding of the complex ways that stress affects communication.

5.1.2. Reduced Emotional and Physical Intimacy

Emotional and physical intimacy in relationships are significantly impacted by stress. Stress frequently causes emotional disengagement and a decrease in emotional reactivity. According to Simpson & Rholes (2017), stressed people may become fixated on their problems, which makes it harder for them to pay attention to and respond to their partner's emotional needs. As a result of this emotional distance, spouses may feel neglected and alone, which can erode their emotional connection (Randall & Bodenmann, 2017).

Stress physiologically triggers the body's stress response system, which releases cortisol and other stress chemicals that might lower arousal and desire for sexual activity (Turner et al., 2020). Prolonged stress can cause exhaustion and a decrease in libido, which can harm a relationship's ability to be physically intimate and satisfying. Stress can also lessen the desire for intimate physical contact and sexual engagement through its physical symptoms, such as tense muscles and sleep disturbances (Simpson & Rholes, 2017).

Furthermore, a feedback loop of emotional and physical disengagement may result from the stress contagion effect, in which the stress of one partner influences the other (Simpson & Rholes, 2017). In order to preserve both emotional and physical closeness in love relationships, managing stress through healthy coping mechanisms and honest communication is essential. Relationship quality and closeness can be markedly improved by interventions that support emotional connection and improve stress management (Randall & Bodenmann, 2017).

5.1.3. Decreased Relationship Satisfaction

In romantic relationships, stress plays a significant role in the reduction of relationship satisfaction. Since stressed people are more likely to be irritable and have negative communication habits, one main mechanism is the frequency and intensity of confrontations increasing (Randall & Bodenmann, 2017). These disputes have the potential to undermine intimacy and trust, which are essential elements of a happy relationship (Kiecolt-Glaser & Wilson, 2017).

Stress frequently results in cognitive biases, which cause people to view their relationships and partners more unfavorably. This negative bias has the potential to skew reality, exaggerating issues and undervaluing the positive features of the relationship (Randall & Bodenmann, 2017). Stress makes these unfavorable opinions worse, which makes partners feel less valued and supported and lowers relationship happiness overall.

Hormonal changes brought on by stress, such as raised cortisol levels, can physiologically impair emotional regulation and increase emotional reactivity. According to Randall & Bodenmann (2017), this increased sensitivity might result in emotional withdrawal or increased conflict, both of which lower relational pleasure. Chronic stress can also result in decreased sexual satisfaction and physical intimacy, which can worsen relationship dissatisfaction.

To sustain and improve relationship happiness, managing stress through practical coping techniques, emotional support, and communication techniques is crucial. Relationship quality and stability can be increased by reducing the negative effects of stress on romantic relationships through interventions that emphasize stress management and emotional connection (Randall & Bodenmann, 2017).

5.1.4. The Stress Contagion Effect

In romantic relationships, stress contagion is the transference of one partner's stress to the other, resulting in a shared sense of anguish. Numerous physiological and psychological mechanisms support this occurrence. Emotional contagion theory, as proposed by Randall & Bodenmann (2017), postulates that people automatically imitate and absorb the emotional states of others who are close to them, especially in close relationships. Stress levels can rise in tandem when one couple is under stress because the other partner may absorb and sense the other's elevated emotional reactivity and bad mood.

From a physiological perspective, stress contagion is caused by the autonomic nerve system being activated. According to research by Monin et al. (2017), during stressful encounters, couples' physiological reactions—such as higher heart rates and cortisol levels—often synchronize. This physiological mirroring makes both partners feel more stressed, which increases the burden on the relationship as a whole.

Stress in a relationship can also spread through behavioral reactions to it, such as irritation, withdrawal, or greater conflict. According to Monin et al. (2017), these unfavorable behaviors may set off a reciprocal cycle of stress reactions in which the stress-related behaviors of each partner intensify and reinforce the stress of the other.

Effective dyadic coping techniques are crucial to reducing the effects of stress contagion. These techniques involve partners managing stress together via communication and support. By strengthening relational resilience and encouraging a more flexible reaction to stress, these strategies safeguard the stability and quality of relationships (Monin et al., 2017).

5.1.5. Increased Conflict

Since stress affects communication and emotional regulation, it is a major cause of conflict in love relationships. According to Randall & Bodenmann (2017), people are more prone to become irritated and have less patience when they are under stress. This might result in more frequent and heated disagreements. Because stress disrupts cognitive processes, it becomes harder to effectively absorb information and react thoughtfully in social situations (Simpson & Rholes, 2017). Hence, miscommunications and intensified arguments may result from partners misinterpreting one another's words and behaviors.

Negative communication patterns that are harmful to relational harmony, such as defensiveness, criticism, and disdain, can be amplified by stress (Kiecolt-Glaser & Wilson, 2017). The probability of conflict is raised by these maladaptive communication practices, and the efficacy of conflict resolution techniques is reduced. Stress-induced emotional reactivity frequently leads to a rapid escalation of arguments because partners may react more strongly to seemingly insignificant concerns.

Stress also reduces empathy and emotional availability, which makes it more difficult for partners to comprehend and accept one another's viewpoints (Kiecolt-Glaser & Wilson, 2017). Ongoing disputes may be stoked by this lack of emotional support, which can intensify feelings of anger and irritation. Reducing the negative effects of stress on interpersonal disputes requires developing better-coping mechanisms, providing emotional support, and practicing effective communication. Stress management interventions can lessen the likelihood that stress will spark conflict and improve marital stability by assisting partners in managing stress more skillfully (Simpson & Rholes, 2017).

5.2. COMMUNICATION STRATEGIES FOR MANAGING STRESS IN RELATIONSHIPS

Scholarly proven communication strategies couples can use to manage stress in relationships are discussed in subsections.

5.2.1. Active Listening

In romantic relationships, active listening is a useful communication technique for stress management since it increases emotional support, develops understanding, and reduces conflict. According to Overall & McNulty (2017), active listening is paying close attention to the speaker, comprehending what they are saying, giving intelligent feedback, and keeping in mind the topic of the conversation. This method is vital for stress reduction because it fosters an atmosphere where both partners feel respected and heard, both of which are critical for emotional health.

According to research, by improving communication quality and avoiding misunderstandings, active listening can dramatically increase relationship satisfaction (Overall & McNulty, 2017). Partners who actively listen to each other are more likely to understand each other's goals and feelings, which reduces the possibility of misunderstandings and conflict. According to Randall & Bodenmann (2017), stress can impair cognitive functioning and increase a person's susceptibility to negative biases and misperceptions, hence it is more vital to accurately understand information during stressful situations.

Additionally, active listening fosters emotional validation and empathy, both of which are essential for offering the emotional support required to manage stress. People can reduce their partner's tension and create an atmosphere of emotional security by recognizing and affirming their partner's sentiments (Overall & McNulty, 2017). In addition to lowering present tension, this sympathetic link fortifies the couple's bond and increases their resistance to stress in the future (Randall & Bodenmann, 2017).

By ensuring that all parties have a thorough awareness of the problems at hand and feel equally involved in finding solutions, active listening also promotes collaborative problem-solving. Effective stress management in relationships depends on a collaborative approach that fosters a sense of teamwork and mutual support (Overall & McNulty, 2017). Additionally, attentive listening may minimize defensive actions and encourage more fruitful communication, which lowers the possibility of conflict and creates a helpful and peaceful interpersonal atmosphere.

5.2.2. Empathetic Communication

Given it promotes emotional connection, understanding, and support, empathic communication is a critical stress management technique in romantic relationships. The capacity to recognize, comprehend, and experience another person's emotions is known as empathy, and it fosters a deep sense of validation and understanding between the two parties (Overall & McNulty, 2017). Empathetic communication in romantic relationships is recognizing and acknowledging one's partner's feelings, listening to them with compassion, and responding in a way that shows concern and understanding.

Empathetic communication is crucial for improving relationship happiness and reducing stress, according to research. Empathetic communication between couples validates one another's experiences and feelings, which has the potential to greatly reduce emotional discomfort. According to Overall & McNulty (2017), this validation lessens the feeling of isolation that frequently accompanies stress by making people feel understood and supported. Additionally, compassionate communication fosters a secure and accepting atmosphere in which partners can freely share their weaknesses and worries without worrying about judgment or rejection (Randall & Bodenmann, 2017).

Further, empathic communication improves emotional regulation, which is essential for stress management. Empathic responses enable partners to support one another in better navigating strong emotions and promote emotional stability. This stability is essential since stress can lead to negative communication patterns and increased emotional reactivity, which can intensify conflicts and strain relationships (Randall & Bodenmann, 2017). On the other hand, sympathetic exchanges can help to diffuse possible disputes and encourage a more positive and encouraging conversation.

More importantly, sympathetic communication improves resilience to stress by fortifying the relationship. Mutual support and cooperative problem-solving are made possible by the emotional intimacy and trust that empathy builds (Overall & McNulty, 2017). As couples who feel connected and supported are better able to confront challenges jointly, this foundation is crucial for managing external stressors (Kiecolt-Glaser & Wilson, 2017).

In relationships, communicating with empathy is a very powerful stress-reduction technique. It is essential for preserving relational stability and health in the face of stress because of its capacity to promote emotional validation, improve emotional regulation, and fortify interpersonal bonds. Couples can manage stress better and keep up a strong, enduring connection by developing empathy (Overall & McNulty, 2017).

5.2.3. Collaborative Problem-Solving

One useful communication technique for reducing stress in romantic relationships, building mutual support, strengthening relational resilience, and encouraging a sense of cooperation is collaborative problem-solving. By collaborating to recognize, manage, and resolve stresses, couples can reduce the detrimental effects of stress on their relationship by fostering a cooperative dynamic (Overall & McNulty, 2017) (Figure 5.2).

Figure 5.2. *Constructive collaborative problem-solving in relationships.*

Source: Shamdani (2022).

Enhancing shared accountability and mutual understanding is a major benefit of collaborative problem-solving. Joint problem-solving encourages partners to talk honestly about their worries and points of view, which helps them come to a more thorough knowledge of the stresses in their lives (Overall & McNulty, 2017). The open lines of communication ensure that each partner feels appreciated and acknowledged, which helps lessen the feelings of annoyance and loneliness that frequently accompany stress (Simpson & Rholes, 2017).

Furthermore, collaborative problem-solving promotes the creation of efficient and amicable solutions. Through combining their knowledge, abilities, and viewpoints, couples can come up with more original and practical stress-reduction plans. By encouraging a sense of teamwork and shared purpose, this cooperative approach not only increases the likelihood of finding successful solutions but also deepens the relationship (Glazer & Liu, 2017).

Collaborative problem-solving not only relieves current pressures but also improves long-term relational resilience. According to research, couples who often work together to solve problems build a stronger foundation for their relationship and a toolkit of efficient coping mechanisms that make them more resilient to stress in the future (Overall & McNulty, 2017). Because couples feel more secure in their ability to overcome obstacles jointly, this increased resilience can result in increased relationship satisfaction and stability (Glazer & Liu, 2017).

Resolving conflicts together can also lessen conflict by reorienting the emphasis from fault finding to solution finding. Partners are less likely to use unfavorable communication styles like criticism or defensiveness when they handle stressors as a team. Alternatively, they could collaborate positively, creating a more uplifting and peaceful interpersonal atmosphere (Overall & McNulty, 2017).

5.2.4. Scheduled Check-Ins

Scheduled check-ins offer a systematic and regular platform for partners to talk about their feelings, thoughts, and pressures, making it an excellent communication technique for stress management in love partnerships. According to Overall & McNulty (2017), this strategy promotes proactive problem-solving, strengthens emotional connection, and keeps lines of communication open.

Scheduled check-ins have the important advantage of giving spouses a set period of time to concentrate on their relationship without outside interruptions. It can be difficult to find time for meaningful conversations amid the hectic pace of everyday life. Frequent check-ins guarantee that both spouses set aside time to prioritize their relationship and deal with problems before they get out of hand (Overall & McNulty, 2017). By doing this, you can stop unresolved stressors from building up, which could otherwise cause conflict and relationship discontent.

Additionally, regular check-ins foster trust and emotional safety. Couples may feel safer and more supported in their relationship when they know they can always count on each other to listen to their worries (Kiecolt-Glaser & Wilson, 2017). This consistency can help people feel less anxious and more stable, which is especially beneficial during trying times. The frequency of these check-ins supports the idea that both parties are dedicated to preserving a happy and caring relationship.

Moreover, scheduled check-ins facilitate proactive problem-solving and collaborative decision-making. By discussing stressors and potential solutions regularly, partners can address issues early and develop effective coping strategies together (Overall & McNulty, 2017). This collaborative approach not only helps in managing immediate stress but also strengthens the relational bond and enhances the couple's resilience to future stressors.

Scheduled check-ins can also improve emotional connection by giving partners a chance to show their gratitude, love, and admiration. According to Randall & Bodenmann (2017), pleasant exchanges during these check-ins can lessen the damaging effects of stress and increase relationship satisfaction overall. An effective stress-reduction strategy can be established by consistently showing support and joy, which will fortify and soothe the connection.

5.3. SUPPORTING A PARTNER THROUGH STRESSFUL TIMES

Supporting a romantic partner through stressful times involves a combination of emotional support, practical assistance, and effective communication. Here are some strategies:

5.3.1. Active Listening

Active listening is a vital strategy for supporting a romantic partner through stressful times, as it fosters emotional connection, reduces misunderstandings, and enhances mutual support. In order to foster a supportive communication environment, active listening entails paying close attention to the speaker, reflecting on what is heard, and demonstrating empathy (Overall & McNulty, 2017).

First, by exhibiting sincere care and empathy, active listening strengthens emotional connection. People who actively listen to their partners communicate that their experiences and feelings are valued, which promotes a feeling of support and affirmation. During trying times, this validation is essential since it reduces feelings of loneliness and fosters emotional health. According to research, having a spouse who understands and supports you can greatly lower stress and increase relationship happiness (Overall & McNulty, 2017).

Misunderstandings and miscommunications are frequent sources of conflict in relationships, especially when things are stressful. Active listening lessens the chance of these occurrences. Active listeners make sure they understand their partner's message correctly by paying close attention to what they are saying and responding in kind. This clarity promotes a more peaceful interpersonal atmosphere and stops stress-related disputes from getting worse (Solomon et al., 2016).

Also, cooperative coping mechanisms and efficient problem-solving are encouraged by active listening. Partners are more willing to have candid and productive conversations about their pressures and potential solutions when they feel heard and understood (Overall & McNulty, 2017). In addition to addressing current stressors, this cooperative strategy improves resilience to upcoming obstacles and fortifies the relationship. By helping to identify particular needs and concerns, active listening enables partners to provide tailored support and collaborate to reduce stress (Randall & Bodenmann, 2017).

5.3.2. Emotional Validation

Supporting a romantic partner through difficult circumstances requires emotional validation, which is identifying and validating their feelings without passing judgment. According to Solomon et al. (2016), this technique can greatly reduce stress and improve the quality of relationships by making people feel understood and accepted.

By recognizing the validity of a partner's emotions and experiences, emotional validation helps to create an atmosphere of psychological safety and emotional support. According to Solomon et al. (2016), couples that affirm each other's emotions convey empathy and understanding, which helps lessen feelings of pain and isolation. This sympathetic reaction is especially important when people are under stress because it reassures them that their feelings are normal and acceptable, which lessens the internal conflict and self-criticism that are frequently linked to stress (Glasdam et al., 2020).

Emotional validation also improves intimacy and communication in relationships. Couples are more likely to communicate their stressors honestly and openly when a nonjudgmental place is provided for them to express their feelings. This promotes understanding between couples and cooperative problem-solving (Overall & McNulty, 2017). This openness can foster a coherent and helpful interpersonal environment and stop stress-related disputes from getting worse.

Since emotional validation strengthens the emotional connection and mutual trust between partners, research suggests that it is linked to higher levels of relationship satisfaction and stability (Solomon et al., 2016). Partners can better manage stress and improve each other's well-being as well as the relationship's general health by validating one another's feelings (Overall & McNulty, 2017).

5.3.3. Open Communication

Due to the fact open communication has such a dramatic effect on relationship dynamics and emotional well-being, it is an essential tactic for supporting a romantic partner through difficult times. Effective communication within relationships functions as a fundamental method for giving emotional and instrumental support, hence mitigating the negative impacts of stressors on individuals, as per the Social Support Theory (Overall & McNulty, 2017). Partners establish a secure space where needs, feelings, and concerns can be communicated without fear of misunderstanding or condemnation by promoting open communication (Solomon et al., 2016).

Furthermore, maintaining relationships through communication is crucial when things are stressful (Solomon et al., 2016). Open communication about stressors helps partners better understand one another's viewpoints and experiences, which promotes intimacy and empathy (Overall & McNulty, 2017). According to Monin et al. (2017), this approach improves relationships and encourages adaptive coping mechanisms including emotional control and problem-solving.

Additionally, Gottman's studies on marriage partnerships highlight the importance of constructive communication, which is defined by validation and active listening, in fostering mutual support and enhancing relationship happiness (Solomon et al., 2016). Thus, encouraging open communication is essential to relational resilience and not only a helpful tactic. It also promotes psychological well-being and flexible coping in stressful situations.

5.3.4. Encouraging Healthy Coping Mechanisms

Based on ideas of stress and coping, encouraging a romantic partner to use healthy coping mechanisms during stressful times is a successful tactic. According to Lazarus and Folkman's Transactional Model of Stress & Coping (1984), people use coping mechanisms to deal with stressors, which affects how they feel and how they think about themselves. Partners can effectively manage stress and lessen its detrimental effects on well-being by encouraging good coping skills like problem-solving, seeking out social support, and practicing relaxation techniques (Overall & McNulty, 2017). Research highlights how adaptive coping can improve resilience and relationship happiness. When it comes to difficult circumstances, couples who encourage one another to employ adaptive coping mechanisms report better relationship quality and less discomfort (Simpson & Rholes, 2017). Furthermore, coping mechanisms that entail interacting with others and taking part in constructive pursuits aid in emotional control and psychological adjustment.

In addition to reducing stress, promoting good coping strategies improves relationships by promoting understanding and support among people. It exhibits compassion and understanding, which are essential for preserving relationship pleasure in the face of difficulty (Solomon et al., 2016). Thus, romantic partners promote each other's resilience and well-being by encouraging adaptive coping, which fosters a strong and successful relationship.

5.3.5. Fostering a Positive Environment

Relationship studies and psychology studies both support the strategy of creating a favorable atmosphere for a love partner during stressful times. Encouragement, optimism, and the establishment of a secure and upbeat environment that protects against the negative impacts of stressors are all components of positive environments. Per the Broaden-and-Build Theory of Positive Emotions (Overall & McNulty, 2017), psychological resilience is strengthened in positive environments because people's short-term thought-action repertoires are expanded, and long-term personal resources are gradually built.

According to research, relationships with good emotional climates tend to be more satisfying and last longer (Solomon et al., 2016). Couples build a foundation of emotional support and understanding by expressing affection, encouragement, laughter, and positivism. Effective coping mechanisms and problem-solving techniques are made easier in this setting, which is essential for handling stressful circumstances (Overall & McNulty, 2017). Furthermore, Gottman's study of marriage partnerships highlights the significance of constructive exchanges, or "bids," which make a substantial contribution to the stability and contentment of a partnership. In addition to intentionally fostering happy and meaningful events, cultivating optimism entails minimizing unpleasant encounters (Overall & McNulty, 2017). Thus, loving couples boost each other's emotional resilience and well-being by creating a good atmosphere, which strengthens and fortifies their bond through trying times.

5.4. MAINTAINING BOUNDARIES IN A RELATIONSHIP

The invisible lines that define our comfort zones are called boundaries. Upholding these limits in a partnership is essential to fostering mutual respect, understanding, and individual well-being. This extensive section examines the value of boundaries, how to set them, and helpful hints for upholding them in a harmonious partnership.

5.4.1. Understanding Boundaries

The restrictions we impose on ourselves to safeguard our mental, emotional, and physical health are known as boundaries. They specify what conduct from other people is appropriate and inappropriate (Gabbard, 2016). Boundaries in a relationship facilitate mutual understanding of needs and respect for personal space. According to Gabbard (2016), boundaries fall into various categories:

1. **Physical Boundaries:** These involve personal space and physical touch. Respecting physical boundaries means understanding how close someone wants to be and what forms of physical contact they are comfortable with.
2. **Emotional Boundaries:** These protect our feelings and emotional health. They involve recognizing and respecting each other's emotions, being mindful of how we express our feelings, and avoiding emotional manipulation or guilt-tripping.
3. **Mental Boundaries:** These pertain to our thoughts, values, and opinions. Respecting mental boundaries means acknowledging and valuing each other's perspectives and not imposing our own beliefs or opinions on our partner.
4. **Sexual Boundaries:** These involve consent, preferences, and privacy in sexual activities. Maintaining sexual boundaries means having open and honest conversations about comfort levels, desires, and limits.
5. **Material Boundaries:** These involve personal possessions and financial resources. Respecting material boundaries means understanding how partners view the sharing of their belongings and finances and agreeing on mutual terms.
6. **Time Boundaries:** These relate to how much time partners spend together and apart. Respecting time boundaries involves balancing time spent with each other and maintaining personal space and activities (Gabbard, 2016).

5.4.2. Importance of Boundaries

Establishing boundaries within a relationship is crucial as it offers multiple benefits, as noted by Joel & MacDonald (2021). The reasons listed below support the idea that maintaining boundaries is essential to a happy and fulfilling relationship.

1. **Encourages Respect:** By recognizing one another's needs and limitations, boundaries encourage respect for one another. The basis of a strong relationship is this respect.
2. **Improves Communication:** Honest and open communication is necessary to establish clear boundaries. Talking about boundaries helps partners communicate their wants and emotions, which improves the bond between them.
3. **Prevents Resentment:** In the absence of limits, couples may feel overburdened or exploited, which breeds animosity. By ensuring that both parties feel appreciated and respected, boundaries help avoid these bad emotions.
4. **Supports Personal Growth:** Boundaries allow individuals to maintain their identity and independence within the relationship. This personal space is crucial for individual growth and self-discovery.
5. **Builds Trust:** When partners respect each other's boundaries, it builds trust. Each partner feels safe and secure, knowing their needs and limits are acknowledged and honored (Joel & MacDonald, 2021).

5.4.3. Establishing Boundaries

Establishing boundaries in a relationship involves several steps. The following are some of the steps an individual can opt for when setting boundaries in their relationship:

1. **Self-Reflection:** Everyone needs time to themselves, whether it be for leisure activities, introspection, or just relaxing. It's critical to talk to your partner about what constitutes appropriate personal space for you both. Respecting your partner's need for private time and space is equally vital, whether that means setting up a specified time each day for personal activities or designating a distinct area in your shared home as "yours."

Tell your partner straight out when you need time alone and invite them to do the same. Setting limits on personal space and time is not about running away from one another; rather, it's about making sure you both get the chance to refuel and develop as people, which in the end makes your relationship stronger.

2. **Communicate Clearly:** Communicating your needs in a relationship is the first step towards establishing boundaries. It's vital to express your feelings openly and honestly, without fear of judgment or criticism. It's not about making demands or ultimatums, but about sharing your feelings, needs, and limits respectfully.

Use "I" statements to avoid blaming or criticizing your partner. For instance, instead of saying "You never spend time with me," say "I feel lonely when we don't spend quality time together." Be as specific as possible and avoid generalizations. Be patient and remember that it might take some time for your partner to understand and respect your boundaries. It's also important to listen to your partner's needs and feelings.

Effective communication is a two-way street. By establishing clear communication, you can create a nurturing environment where both of you feel heard, understood, and valued.

3. **Mutual Agreement:** Understanding and respecting your partner's boundaries is equally important as establishing your own. This means acknowledging their limits, even if they are different from yours. Show them you respect their boundaries by always asking for their consent, not pressuring them to change their mind, and apologizing when you accidentally cross a line.

It's crucial to create an environment where both parties feel comfortable expressing their needs without fear of retaliation. Make it a point to have regular conversations about what is and isn't working. Open communication is key to understanding and upholding boundaries. While the process may be challenging, the reward is a relationship built on mutual respect and trust. Remember, maintaining boundaries isn't about restrictions or limitations; it's about respect and understanding.

4. **Be Consistent:** Consistency is key to maintaining boundaries. Stick to your boundaries and reinforce them when necessary. Also, boundaries may change over time. Regularly revisit and adjust them as needed. Open communication about any changes ensures both partners remain on the same page (Gabbard, 2016).

5.4.4. Keeping Boundaries in Place

Setting and upholding limits takes constant work and dedication. Here are a few useful pointers:

1. **Acknowledge Each Other's Personal Space:** Give your partner time and space to themselves. Honor their desire for privacy or the things they like to accomplish by themselves.
2. **Actively Listen:** Be mindful of your partner's wants and emotions. You can better understand and acknowledge their boundaries if you actively listen to them.
3. **Don't Assume:** Don't presume to understand your partner's wants or emotions. Always solicit their opinions and show respect for their answers.

4. **Be Honest:** When a boundary has been crossed, confront it calmly and sincerely. Talk about what occurred, how it affected you, and how to avoid it going forth.
5. **Practice Self-Care:** Taking care of your own physical, emotional, and mental well-being is crucial for maintaining boundaries. Engage in activities that rejuvenate and energize you.
6. **Seek Support:** If maintaining boundaries becomes challenging, seek support from a therapist or counselor. Professional guidance can provide tools and strategies to reinforce boundaries effectively (Gabbard, 2016).

5.4.5. Common Challenges and Solutions

Maintaining boundaries can be challenging, especially in close relationships. Here are some common challenges an individual might face in keeping boundaries and solutions to the given situations:

1. **Guilt:** Feeling guilty for setting boundaries is common. Remember, boundaries are necessary for your well-being. Communicate to your partner that your boundaries are not a rejection of them but a way to take care of yourself.
2. **Resistance from Partner:** Your partner may resist your boundaries initially. Be patient and explain why these boundaries are important to you. Encourage open dialogue to address any concerns.
3. **Overstepping Boundaries:** Sometimes, boundaries may be unintentionally crossed. When this happens, address it immediately. Reiterate your boundaries and discuss ways to prevent future occurrences.
4. **Cultural Differences:** Cultural backgrounds can influence perceptions of boundaries. Have open conversations about these differences and find common ground that respects both partners' values and traditions.

5.5. SEEKING COUPLE THERAPY FOR STRESS

In the modern world, stress is an unavoidable part of life that affects not only people but also their relationships. Stress may be a big problem for couples, possibly weakening their relationship and causing arguments. Without outside assistance, managing relationship tension can occasionally be challenging. A certified couple's therapist can be a beneficial resource since they typically possess the information, and abilities required to assist you in lowering your stress level and strengthening your bond. To address and manage these problems and promote a more robust relationship, seeking couple therapy for stress might be a useful strategy (Figure 5.3).

Figure 5.3. *Couples therapy.*

Source: Goltzman (2023).

5.5.1. A Brief Overview of How Stress Affects Relationships

Numerous factors can lead to stress, such as obligations to one's family, work pressures, money problems, and health problems (Randall & Bodenmann, 2017). Stress in one or both spouses can show up in the relationship in several ways:

1. **Communication Breakdown:** Stress can make it difficult to communicate effectively, which can result in fights, misunderstandings, and a lack of emotional bond.
2. **Emotional Distancing:** When a couple is emotionally detached from one another, they may feel less affectionate and encouraging towards one another.
3. **Enhanced Conflict:** Stress frequently intensifies conflict and makes small problems appear intractable.
4. **Decreased Intimacy:** Physical closeness can be impacted by high levels of stress, which can lower sexual desire and fulfillment (Randall & Bodenmann, 2017).

5.5.2. The Role of Couple Therapy in Managing Stress

Couple therapy is a sort of psychotherapy intended to assist couples in strengthening their bond and resolving difficulties. It is often referred to as marriage counseling or couples counseling (Parnell & Hammer, 2018). Couples therapy has numerous important advantages for stress management which entails the following:

1. **Better Communication:** Couples who receive good communication skills from therapists are able to communicate more honestly and openly about their thoughts and concerns.
2. **Conflict Resolution:** Counselling offers techniques for settling disputes amicably, assisting partners in reaching a compromise and a common ground.
3. **Emotional Assist:** Couples get stronger in their relationship and increase their ability to withstand stress by learning how to better assist one another emotionally.
4. **Stress Management Skills:** Couples can practice stress management skills that therapists provide, such as mindfulness exercises, relaxation techniques, and coping mechanisms together (Parnell & Hammer, 2018).

5.5.3. The Couples Therapy Process

There are various steps involved in seeking couple therapy for stress, each aimed at addressing a particular facet of the relationship and the unique stresses that impact it (Parnell & Hammer, 2018):

1. **First Evaluation:** The first assessment is the first stage of couple therapy, during which the therapist meets with the couple to learn about their concerns, relationship background, and current stressors. This meeting aids the therapist in customizing treatment to meet the specific needs of the couple.
2. **Establishing Objectives:** Following the evaluation, the therapist collaborates with the couple to establish clear, attainable goals for their therapy. These objectives could be learning stress-reduction strategies, settling persistent disputes, or enhancing communication.
3. **Developing a Treatment Plan:** Based on the assessment and goals, the therapist develops a customized treatment plan. This plan outlines the therapeutic approaches and techniques that will be used, the frequency of sessions, and any homework assignments for the couple.

4. **Therapy Sessions:** Typically, therapy sessions consist of both individual and group sessions. In these sessions, the therapist assists the couple in exploring their emotions, recognizing harmful habits, and putting new skills into practice. Typical treatment modalities consist of:
    i. **Cognitive-Behavioral Therapy (CBT):** This type of therapy assists couples in recognizing and altering harmful thought patterns and behaviors that fuel tension and conflict.
    ii. **Emotionally Focused Therapy (EFT):** This aims to strengthen partners' emotional bond and attachment.
    iii. **The Gottman Method:** This strategy places a focus on developing a close friendship, handling disagreements, and giving the partnership a shared purpose (Parnell & Hammer, 2018).
5. **Assignments and Practice:** As a means of encouraging couples to practice new skills and approaches in between sessions, homework assignments are frequently assigned in couple therapy. These tasks could involve bond-building activities, stress-reduction techniques, and communication exercises.
6. **Assessment of Progress:** The therapist periodically assesses the couple's advancement towards their objectives during the course of therapy. To make sure the therapy continues to be effective, changes to the treatment plan may be necessary (Lebow & Snyder, 2022).

5.5.4. Couples Therapy's Benefits for Stress

Couples therapy for stress reduction has several advantages for the partners as well as the relationship itself. They include:

1. **Stronger Relationship:** Therapy helps couples build stronger, more resilient relationships by improving communication, emotional support, and conflict resolution skills.
2. **Enhanced Emotional Well-Being:** Reducing stress and improving the relationship can lead to better emotional well-being for both partners.
3. **Improved Physical Health:** Chronic stress can take a toll on physical health. By managing stress through therapy, couples may experience improved overall health.
4. **Greater Life Satisfaction:** A healthy relationship contributes significantly to overall life satisfaction and happiness.

5.5.5. When to Seek Stress Couples Therapy?

Knowing when to go to marital therapy is essential to managing stress before it seriously damages the relationship. Here are a few indicators that it might be time to start therapy:

1. **Persistent Arguments:** Recurrent, unresolved disputes that appear to get worse instead of better.
2. **Emotional Distance:** Sensing a lack of emotional connection with your significant other.
3. **Communication Problems:** Having trouble understanding one another or expressing one's thoughts and feelings.
4. **Decreased Intimacy:** There has been a discernible loss of both emotional and physical intimacy.
5. **Incapacity to Handle Stress:** Having trouble controlling stress and how it affects the relationship (Lebow & Snyder, 2022).

5.5.6. Selecting the Appropriate Therapist

The success of marital therapy depends on selecting the appropriate therapist. When choosing a therapist, take into account the following aspects:

1. **Qualifications and Experience:** Look for a licensed therapist with experience in couple therapy and a background in dealing with stress-related issues.
2. **Approach and Techniques:** Ensure the therapist's approach aligns with your needs and preferences.
3. **Comfort and Trust:** It's important to feel comfortable and build trust with your therapist. A good rapport can enhance the therapeutic process.

5.6. BALANCING INDIVIDUAL AND RELATIONSHIP NEEDS

A vital component of any successful partnership is striking a balance between individual and interpersonal requirements. Humans naturally want to connect with other people, but we also have wants of our own that must be satisfied to stay healthy. Although it can be difficult, doing this is necessary to build a happy and satisfying relationship.

Finding balance begins with acknowledging one's own needs as well as those of relationships. Relationship requirements might include closeness, communication, and trust, while personal needs can include mental, emotional, and physical health. It's critical to understand that relationship needs, and personal needs are interdependent, and that ignoring one might result in an unhappy partnership (Boszormenyi-Nagy, 2014).

5.6.1. Recognizing Personal Needs

Personal needs include all of a person's wants and requirements that are essential to their physical, mental, and emotional well. These requirements can be roughly divided into:

i. **Physical Needs:** Access to proper sleep, food, exercise, and medical care.

ii. **Emotional Needs:** A sense of security, understanding, love, and worth.

iii. **Needs in Terms of Psychology:** Competence, autonomy, and purpose.

iv. **Social Needs:** A sense of community, belonging, and outside assistance.

People may find it difficult to meet these demands when they are under stress. Prolonged stress can cause mental health disorders like anxiety and depression as well as physical health problems like hypertension and compromised immune systems. Stress can also make it harder for a person to meet their psychological and emotional requirements, which can leave them feeling overwhelmed and less satisfied with their lives (Randall & Bodenmann, 2017).

5.6.2. Recognizing Relationship Requirements

Relationship needs are the prerequisites necessary to preserve wholesome, satisfying interpersonal relationships. Important relationship requirements consist of:

1. **Communication:** The free, courteous, and sincere sharing of ideas and emotions.
2. **Trust:** The ability to be dependable, truthful, and faithful in a partnership.
3. **Support:** A partner's practical, psychological, and emotional help.
4. **Intimacy:** Sharing experiences, being physically near, and having an emotional bond.
5. **Conflict Resolution:** Skillfully handling disputes and divergences.

These relationship-related factors can be greatly impacted by stress. For instance, a high stress level might result in decreased closeness, poor communication, and more conflict. Furthermore, the ability of one or both partners to support and uphold trust may be jeopardized when they are under stress (Randall & Bodenmann, 2017).

5.6.3. Balancing Individual and Relationship Needs

Balancing individual and relationship needs in the context of stress involves recognizing the importance of both sets of needs and finding strategies to address them simultaneously (DeJonckheere & Vaughn, 2019). Here are key approaches to achieving this balance:

1. **Awareness of Oneself and Self-Care:** The cornerstone of stress management and need balancing is self-awareness. People need to be aware of the things that stress them out and know how they respond to stress. Taking care of oneself is essential to preserving one's well-being (Boszormenyi-Nagy, 2014).

Self-care may consist of:

- i. **Physical Self-Care:** This includes regular exercise, a balanced diet, getting enough sleep, and visiting the doctor.
- ii. **Emotional Self-Care:** Pastimes, meditation, and quality time with loved ones are examples of activities that encourage contentment and relaxation.
- iii. **Mental Self-Care:** Mentally engaging activities that promote a feeling of accomplishment, such as reading, picking up new skills, or engaging in artistic pursuits.
- iv. **Making Self-Care a Priority:** This helps people satisfy their needs and better manage stress, which in turn makes it possible for them to positively impact relationships.

2. **Effective Communication:** Honesty and clarity in communication are crucial in partnerships. Building trust and avoiding misunderstandings can be achieved by being truthful with our spouse. Open and honest communication of our wants and desires is crucial, even though it might be challenging or uncomfortable at times. This can foster a closer emotional bond and help avoid resentment.

Additionally crucial to communication is clarity. We should make an effort to speak clearly and stay away from unclear or imprecise words. By doing this, we can make sure that our partner is aware of our requirements and preferences and avoid misunderstandings. We ought to be receptive to criticism and prepared to dispel any misconceptions (Boszormenyi-Nagy, 2014).

Effective communication is essential to balancing personal and relationship needs in a healthy relationship. By actively listening and communicating honestly and clearly, we can build trust, strengthen emotional connections, and prevent misunderstandings.

3. **Supporting Each Other:** In a relationship, mutual support is crucial for stress management. During difficult circumstances, partners should make an effort to:
   i. **Provide Emotional Support:** Empathy, validation, and reassurance.
   ii. **Share Duties:** To lessen stress for both partners, divide up domestic chores and duties.
   iii. **Promote Self-Care:** Show respect for one another's personal space and time, as well as support one another's self-care routines.

Strengthening the relationship and improving both partners' ability to handle stress can be achieved via mutual understanding and support of one another's needs.

4. **Setting Boundaries:** It is crucial for maintaining a balance between individual and relationship needs. Healthy boundaries can include:
   i. **Personal Time:** Allocating time for individual activities and interests without feeling guilty or neglecting the relationship.
   ii. **Emotional Boundaries:** Recognizing when to step back and give each other space to process emotions and stress independently.
   iii. **Relationship Boundaries:** Establishing limits on behaviors that can increase stress, such as excessive criticism or unrealistic expectations.

Boundaries help ensure that both partners can meet their individual needs while maintaining a healthy, supportive relationship.

5. **Resolving Conflicts:** Relationship issues can be made worse by stress, which is why it's so important to resolve conflicts amicably. Important tactics consist of:
   i. **Remain Calm:** Control one's tension and feelings before intervening in disputes.
   ii. **Keep Your Attention on the Issue:** Dealing with the current issue at hand rather than bringing up old grudges.
   iii. **Seek Compromise:** Coming up with solutions that take into account the requirements and viewpoints of both parties and are mutually agreeable (DeJonckheere & Vaughn, 2019).

Resolving conflicts effectively eases tension for both parties and preserves the harmony of the partnership.

6. **Getting Expert Assistance:** Seeking professional assistance might be helpful when stress gets too much and interferes with one's ability to function both personally and in relationships. Choices consist of:
   i. **Individual Therapy:** Developing coping mechanisms and addressing personal concerns in collaboration with a therapist.
   ii. **Couples Therapy:** Seeing a therapist together to develop your relationship, resolve disputes, and enhance communication.
   iii. **Support Groups:** Attending events where singles and couples can talk to others going through similar struggles about their experiences and coping mechanisms.

Insights and useful tools for stress management and juggling the demands of relationships and oneself can be obtained with professional assistance (DeJonckheere & Vaughn, 2019).

We have explored the importance of balancing personal and relationship needs. We have seen that it is essential to maintain a sense of self while being in a relationship. It is important to communicate with our partners and discuss our individual needs and relationship goals. By doing so, we can create a healthy and fulfilling relationship.

5.7. BUILDING RESILIENCE IN RELATIONSHIPS

Relationships are always put to the test by many circumstances, whether they are romantic, familial, or platonic. Stress can seriously damage relationships due to a variety of factors, including health crises, work-related difficulties, and financial hardship. Developing relationship resilience is essential to withstanding these demands and preserving a strong, sustaining bond. Although there is no road map for life, everyone will encounter obstacles along the way, from little setbacks to major life-changing incidents like a loved one passing away, a catastrophic accident, or a terrible disease. Every shift has a particular impact on each individual, bringing with it a rush of ideas, intense feelings, and uncertainties. However, resilience plays a part in people's ability to adjust well over time to difficult and life-altering circumstances.

Psychologists define resilience as the process of adapting well in the face of adversity, trauma, tragedy, threats, or significant sources of stress—such as family and relationship problems, serious health problems, or workplace and financial stressors. As much as resilience involves "bouncing back" from these difficult experiences, it can also involve profound personal growth.

Even while these unfavorable circumstances are painful and challenging, much like rough river waves, they don't have to dictate how your life turns out. You can change, adapt, and progress in many areas of your life. That is resilience's function. Developing greater resilience gives you the ability to not only overcome adversity but also to advance and even enhance your life in the process.

Resilience in a relationship refers to a partner's capacity to overcome obstacles, adjust to changing circumstances, and keep a positive, working relationship in the face of hardship. Effective communication, mutual support, and a shared dedication to conquering challenges are traits of resilient relationships.

5.7.1. Crucial Techniques for Developing Resilience

1. **Effective Communication:** The foundation of any strong relationship is communication. Communicating in an open, sincere, and compassionate manner enables partners to share their needs, wants, and worries. It promotes comprehension and averts misconceptions that may increase tension.
   i. **Active Listening:** It requires bravery to remain silent while your significant other is discussing something. While you might feel compelled to correct them or clarify yourself, spending time listening to them will improve your relationship.
   ii. **Expressing Emotions:** Encourage each other to express emotions freely. Use "I" statements to communicate feelings without blaming or criticizing. For example, say, "I feel overwhelmed when…" instead of "You always make me feel…"
   iii. **Regular Check-Ins:** Schedule regular check-ins to discuss stressors and emotional well-being. This practice helps partners stay connected and address issues before they escalate (Forbes & Fikretoglu, 2018).
2. **Sharing Empathy and Support:** Empathy and mutual support are essential for developing resilience. Assisting one another during trying times makes the relationship stronger and promotes a sense of cooperation (McCraty & Atkinson, 2012).
   i. **Emotional Support:** Reassure your partner and validate their feelings to provide emotional support. Acknowledge their anxiety and let them know you're eager to support them.
   ii. **Practical Support:** Assist others practically by allocating duties and responsibilities. For example, offer to help with more housework or run errands if your partner is overworked and stressed.

iii. **Empathy:** Put yourself in your partner's position to develop empathy. Recognize their viewpoint and act compassionately in response. Empathy fosters a sense of unity and lessens emotions of isolation.

3. **Stress Management Techniques:** By implementing stress-reduction strategies into your partnership, you may improve stress management for both parties.

   i. **Meditation and Mindfulness:** Combine mindfulness and meditation exercises. These methods enhance emotional control, induce relaxation, and lessen worry.

   ii. **Exercise:** Take part in physical exercises like yoga, jogging, or strolling with your partner. Exercise releases endorphins, which have been shown to elevate mood and lower stress levels.

   iii. **Interests and Hobbies:** Follow your interests and hobbies jointly. Fun activities help people decompress and improve their emotional relationships (Forbes & Fikretoglu, 2018).

4. **Problem-Solving Skills:** Effective problem-solving skills are essential for navigating stressors and preventing conflicts from escalating.

   i. **Identify the Problem:** Identify the source of stress and agree on the problem that needs to be addressed.

   ii. **Brainstorm Solutions:** Collaboratively brainstorm possible solutions. Encourage creativity and consider various options without judgment.

   iii. **Evaluate and Choose:** Evaluate the potential solutions and choose the one that is most feasible and acceptable to both partners.

   iv. **Implement and Review:** Implement the chosen solution and review its effectiveness. Be flexible and willing to make adjustments as needed (Forbes & Fikretoglu, 2018).

5. **Building a Support Network:** A strong support network outside the relationship can provide additional resources and reduce the burden on partners.

   i. **Family and Friends:** Maintain connections with family and friends who can offer support, advice, and a listening ear.

   ii. **Professional Help:** Seek professional help, such as counseling or therapy, when needed. A therapist can provide tools and strategies to manage stress and improve relationship dynamics.

iii. **Community Involvement:** Engage in community activities and groups that offer support and foster a sense of belonging.

6. **Positive Relationship Practices:** Positive relationship techniques can strengthen resilience and mitigate the damaging consequences of stress.
   i. **Gratitude:** Show appreciation for one another regularly to cultivate gratitude. Celebrate and acknowledge each other's accomplishments.
   ii. **Affection:** Express your affection with loving gestures, words of kindness, and physical contact. Oxytocin, which is released during physical affection, fosters bonding and lowers stress.
   iii. **Quality Time:** Make the most of your time together by avoiding distractions. Take part in enjoyable activities that enhance the emotional bond between partners (Mansfield et al., 2016).
7. **Flexibility and Adaptability:** Resilience requires the ability to be adaptive and flexible in the face of change. Since life is unpredictable, being able to adapt to changing circumstances is a useful skill.
   i. **Acceptance:** Accept that change is a natural part of life and that some stressors are beyond your control. Focus on what you can control and adapt to new situations.
   ii. **Flexibility:** Be willing to adjust your expectations and plans. Flexibility allows you to navigate challenges more smoothly and reduces the strain on the relationship.
   iii. **Growth Mindset:** Adopt a growth mindset, viewing challenges as opportunities for growth and learning. A positive outlook can enhance resilience and promote a sense of purpose (Forbes & Fikretoglu, 2018).

CHAPTER 6

STRESS AND AGING

Contents

The scientific community has given great attention to the relationship between stress and aging due to its important implications for both individual and public health. The effects of stress, a complex physiological and psychological reaction to external demands, on a range of biological systems have been well researched. Numerous genetic, environmental, and lifestyle factors can have an impact on aging, which is defined by a steady deterioration in physiological capabilities and an increasing vulnerability to diseases. Significant research highlights the harmful consequences of stress on cellular, systemic, and psychological levels. An increasing body of evidence clarifies the complex link between chronic stress and accelerated aging (Cohen et al., 2016) (Figure 6.1).

Figure 6.1. *The link between stress and aging.*

Source: Longevity (2021).

Chronic stress has been linked to accelerated cellular aging processes at the cellular level, primarily via telomere degradation, the eroding of chromosomal ends that serve as protective caps. Higher stress levels have been associated with telomere shortening, a hallmark of cellular aging, indicating that cellular senescence accelerates in these individuals (Birch-Machin & Bowman, 2016). Furthermore, a major factor in driving the aging process is the dysregulation of the hypothalamic-pituitary-adrenal (HPA) axis brought on by stress. Chronic HPA axis activation raises cortisol levels, which can lead to oxidative stress, inflammatory pathways, and compromised immunological function—all of which are linked to age-related diseases (Amarya et al., 2018).

Systemically, chronic stress has a role in the development and advancement of age-related illnesses such as diabetes, neurodegenerative diseases, and cardiovascular disease. The cumulative effect of prolonged exposure to stress, also known as "allostatic load," is the wear and tear that the body experiences over time, exacerbating the deterioration of physiological resilience and functional performance (Amarya et al., 2018). Stress can negatively impact mental health psychologically, resulting in disorders like anxiety and depression that worsen the quality of life and exacerbate the aging process (Cohen et al., 2016).

Understanding the mechanisms through which stress influences aging is paramount for developing targeted interventions aimed at mitigating its adverse effects. Strategies such as stress management, lifestyle modifications, and pharmacological approaches hold promise in promoting healthy aging and enhancing overall longevity (McEwen, 2017).

Going forward, this chapter delves into the impacts of chronic stress on the aging process, and the stress management options at old age.

6.1. EFFECTS OF STRESS ON AGING

The effects of stress on aging are profound and multifaceted, encompassing cellular, systemic, and psychological dimensions. Some effects of stress on aging supported by scholarly sources are as in subsections.

6.1.1. Dysregulation of the Hypothalamic-Pituitary-Adrenal (HPA) Axis

The HPA axis is crucial for the body's reaction to stress as it controls the adrenal cortex's secretion of glucocorticoids, mainly cortisol. Prolonged exposure to increased cortisol levels is caused by the HPA axis being persistently activated by chronic stress. Through several pathways, this disruption of the HPA axis greatly accelerates the aging process (Yaribeygi et al., 2017).

Firstly, sustained high cortisol levels negatively impact the immune system, reducing its efficiency and leading to increased susceptibility to infections and slower wound healing (McEwen, 2017). Chronic elevation of cortisol impairs the production and function of immune cells, such as lymphocytes, and promotes a pro-inflammatory state. Inflammation, in turn, is a critical driver of aging, contributing to the development of age-related diseases, including cardiovascular disease, diabetes, and neurodegenerative disorders (Yaribeygi et al., 2017).

Second, cortisol negatively impacts the cardiovascular system. The metabolic syndrome, hypertension, and atherosclerosis are linked to chronic activation of the HPA axis. High cortisol levels increase blood pressure and changes lipid metabolism, both of which promote arterial plaque buildup and blood vessel stiffness. According to McEwen (2017), these cardiovascular alterations are characteristic of aging and greatly increase the risk of myocardial infarction and stroke.

Moreover, long-term exposure to cortisol damages brain function, especially in areas of the brain linked to memory and cognition, such as the hippocampus. According to Yaribeygi et al. (2017), high cortisol levels cause hippocampus atrophy, decrease neurogenesis, and synaptic plasticity, all of which are linked to cognitive decline and an increased risk of dementia in older persons. The fact that cortisol can worsen oxidative stress and encourage neuronal death adds to its neurotoxic effects. Finally, dysregulation of the HPA axis and chronic stress impact metabolic processes, resulting in insulin resistance, belly fat, and muscular atrophy. These aging-related metabolic abnormalities lead to diminished physical function and frailty (McEwen, 2017).

6.1.2. Oxidative Stress and Inflammation

Chronic stress accelerates the aging process through two key mechanisms: oxidative stress and inflammation. Chronic stress physiologically promotes oxidative stress and inflammation, and these two factors greatly accelerate cellular and systemic aging (Amarya et al., 2018).

An imbalance between the body's antioxidant defense mechanisms and the generation of reactive oxygen species (ROS) leads to oxidative stress. Prolonged stress increases the generation of reactive oxygen species (ROS) by a number of mechanisms, including mitochondrial malfunction and the activation of enzymes like NADPH oxidase. The cellular constituents, such as DNA, proteins, and lipids, are severely harmed by these ROS (Birch-Machin & Bowman, 2016). Mutations and genomic instability are characteristic elements of aging and age-related disorders. DNA damage, especially in the form of double-strand breaks and oxidative base alterations, can result in these conditions (Amarya et al., 2018). Apoptosis and cellular senescence are facilitated by protein oxidation and lipid peroxidation, which compromise cellular integrity and function.

Inflammation, another critical aspect of chronic stress, is closely intertwined with oxidative stress. Chronic stress activates the HPA axis and sympathetic nervous system, leading to increased levels of glucocorticoids and catecholamines. These stress hormones modulate the immune response, promoting a pro-inflammatory state characterized by elevated levels of inflammatory cytokines such as interleukin-6 (IL-6) and tumor necrosis factor-alpha (TNF-α) (Liu et al., 2017). Persistent inflammation, or inflammaging, contributes to the aging process by exacerbating tissue damage and impairing tissue repair mechanisms.

Numerous age-related illnesses show the complementary impacts of inflammation and oxidative stress. According to Birch-Machin & Bowman

(2016), oxidative and inflammatory damage to endothelial cells causes atherosclerosis, which in turn causes cardiovascular disease, a major cause of morbidity and mortality in the aged. Likewise, oxidative stress-induced neuronal damage and persistent neuroinflammation are linked to neurodegenerative illnesses like Alzheimer's disease, hastening the deterioration of cognitive function (McEwen, 2017). Furthermore, inflammation and oxidative stress compromise the functions of autophagy and proteostasis, two cellular repair systems. Chronic inflammation inhibits autophagy, a process necessary for the elimination of damaged proteins and organelles. This results in the buildup of cellular waste and accelerates the aging process of cells (Amarya et al., 2018). Protein homeostasis, or proteostasis, is upset, leading to an accumulation of damaged and misfolded proteins that are hazardous to cells and encourage aging-related illnesses.

6.1.3. Telomere Shortening

Chronic stress has a major impact on telomere shortening, a hallmark of biological aging that accelerates the aging process in both cells and organisms. Telomeres are chromosomal ends that contain repeating nucleotide sequences that shield the ends from deterioration and stop important genetic information from being lost during cell division. Telomeres gradually shorten with each cell division, and when severely short telomeres cause DNA damage responses, cellular senescence, or apoptosis results (de Magalhães & Passos, 2018) (Figure 6.2).

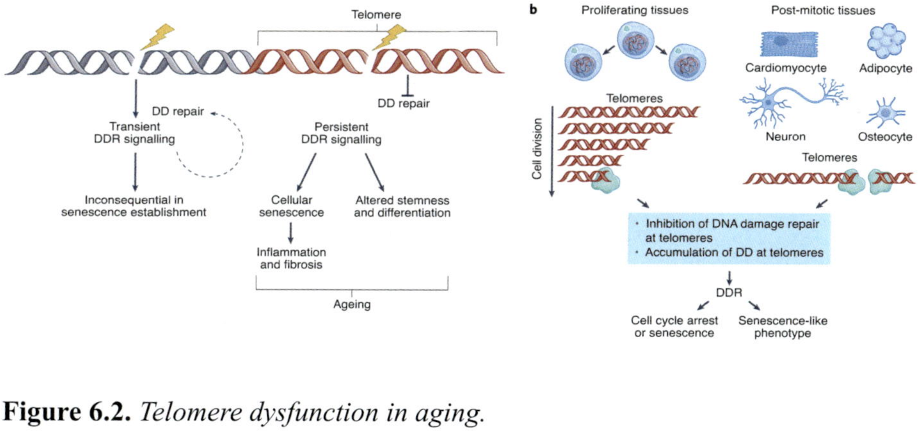

Figure 6.2. *Telomere dysfunction in aging.*

Source: Rossiello et al. (2022).

Telomere shortening is accelerated by chronic stress via a number of biological pathways. One main mechanism is the chronic activation of the HPA axis, which raises cortisol levels. Because telomeric DNA has a higher guanine content than other genomic areas, damage from ROS is more likely to occur when cortisol levels are high. This can lead to oxidative stress. (Amarya et al., 2018). Telomere attrition is made worse by this oxidative damage in addition to the typical replicative shortening that happens during cell division.

Moreover, stress-induced behavioral factors such as poor sleep, unhealthy diet, and lack of physical activity further contribute to telomere shortening. These lifestyle factors can enhance systemic inflammation and oxidative stress, thereby accelerating cellular aging. For instance, inflammation activates immune cells that produce ROS and secrete inflammatory cytokines, promoting a pro-oxidative environment detrimental to telomere integrity (de Magalhães & Passos, 2018).

The aging process and age-related illnesses are significantly impacted by telomere shortening. Cellular senescence, an irreversible cell cycle arrest state that hinders tissue regeneration and function, is brought on by critically short telomeres. The senescence-associated secretory phenotype (SASP), which is secreted by senescent cells, is linked to tissue dysfunction and chronic inflammation. It consists of pro-inflammatory cytokines, chemokines, and proteases (Birch-Machin & Bowman, 2016). This pro-inflammatory milieu accelerates the aging process and fosters the emergence of age-related illnesses like cancer, neurological diseases, and cardiovascular disease.

Furthermore, telomere shortening in stem cells impairs their regenerative capacity, leading to the decline of tissue homeostasis and function. For example, in hematopoietic stem cells, telomere attrition reduces their ability to produce new blood cells, contributing to immunosenescence and increased susceptibility to infections (Amarya et al., 2018).

6.1.4. Psychological Morbidity

Through a variety of neurobiological and psychosocial mechanisms, chronic stress plays a crucial role in the development of psychological morbidity during the aging process. According to McEwen (2017), stress-related psychiatric illnesses like depression and anxiety in older individuals arise as a result of the chronic activation of the stress response system, which includes the sympathetic nervous system and the HPA axis.

The disruption of the HPA axis is one of the main neurobiological routes by which long-term stress causes mental health issues. Extended periods of stress result in persistently high levels of glucocorticoids, especially cortisol. The hippocampus, prefrontal cortex, and amygdala are three important brain areas involved in mood regulation that are impaired in chronic hypercortisolemia. The hippocampus, which is important for controlling emotions and memory, is especially susceptible to harm from cortisol. According to McEwen (2017), long-term stress causes hippocampus atrophy, decreased neurogenesis, and altered synaptic plasticity, all of which are linked to cognitive impairment and a higher risk of depression.

Similarly, the prefrontal cortex, responsible for executive functions and emotional control, exhibits structural and functional deficits under chronic stress, further exacerbating depressive and anxiety symptoms (Amarya et al., 2018).

Apart from causing changes in the nervous system, long-term stress also creates psychosocial problems that lead to mental illness. Birch-Machin & Bowman (2016) found that depression and anxiety in older persons are significantly influenced by stressful life events and long-term psychosocial stressors, such as providing care for a spouse or facing financial challenges. A high correlation has been shown between psychological distress and mental health issues and the outcomes of these stressors, which include social isolation, diminished social support, and helplessness.

Furthermore, it is becoming more widely acknowledged that inflammation is a key mediator of depression and that persistent stress increases it. Chronic stress and depression have been linked to elevated levels of pro-inflammatory cytokines, including tumor necrosis factor-alpha (TNF-α) and interleukin-6 (IL-6) (de Magalhães & Passos, 2018). By changing neurotransmitter metabolism, decreasing neuroplasticity, and rupturing the blood-brain barrier, these cytokines can impact brain function and play a role in the onset and maintenance of depression symptoms.

The cumulative effects of these neurobiological and psychosocial mechanisms highlight the profound impact of chronic stress on psychological morbidity in the aging population. Effective stress management and interventions targeting these pathways are essential for improving mental health and quality of life among older adults (Amarya et al., 2018).

6.2. MANAGING STRESS IN OLDER ADULTS

Effective stress management is crucial for promoting mental and physical health in the older adult population. Common evidence-based strategies for managing stress in this demographic, supported by scholarly sources include the following:

6.2.1. Mindfulness and Meditation Practices

Meditating and practicing mindfulness have become popular stress-reduction strategies for older adults, with significant psychological and physical advantages. These techniques entail bringing one's attention to the here and now while embracing and nonjudgmentally addressing one's feelings and ideas.

Programs such as Mindfulness-Based Stress Reduction (MBSR) aim to develop these abilities, resulting in significant decreases in stress and enhancements in general well-being (Geiger et al., 2016).

The modulation of the HPA axis is one of the main ways by which mindfulness and meditation reduce stress. Prolonged stress triggers the HPA axis, raising cortisol levels that may be harmful to one's health. It has been demonstrated that mindfulness techniques lower cortisol levels, which lessens the physiological effects of stress (Khoury et al., 2015). Lower levels of anxiety and sadness, two prevalent stress-related illnesses in older persons, are linked to lower cortisol levels.

Furthermore, through raising activity in the prefrontal cortex—a part of the brain linked to executive functions and emotional control—mindfulness and meditation improve emotional regulation. By improving their ability to respond to stimuli, people are less likely to have unfavorable emotional reactions thanks to this increased activity (Sharma & Rush, 2014). Additionally, mindfulness exercises enhance cognitive flexibility and attentiveness, both of which are critical for adjusting to the difficulties of aging.

Empirical studies have demonstrated the efficacy of mindfulness and meditation in improving mental health outcomes among older adults. For instance, Geiger et al. (2016) found that MBSR programs significantly reduced perceived stress and enhanced quality of life in elderly participants. These practices also promote social connectedness and support, further buffering against the adverse effects of stress.

6.2.2. Cognitive Behavioral Therapy

As a highly successful method of treating both the behavioral and cognitive components of stress as well as the associated psychiatric problems, cognitive behavioral therapy (CBT) is used with older persons. According to Khoury et al. (2015), this therapy approach is predicated on the idea that maladaptive thought patterns and beliefs can be reorganized to enhance mental health outcomes by contributing to emotional suffering and behavioral problems (Figure 6.3).

Cognitive behavioral therapy (CBT) helps older persons recognize and address stress-inducing negative thoughts and cognitive distortions. Through the process of cognitive restructuring, people are able to replace their stressful and illogical thoughts with more realistic and balanced ones. This change in thinking lessens the severity and occurrence of stress reactions (Yaribeygi et al., 2017). An older adult, for instance, may overstress health problems, thinking that even slight symptoms point to a catastrophic condition. CBT assists in reframing these kinds of ideas, which lowers anxiety and related stress.

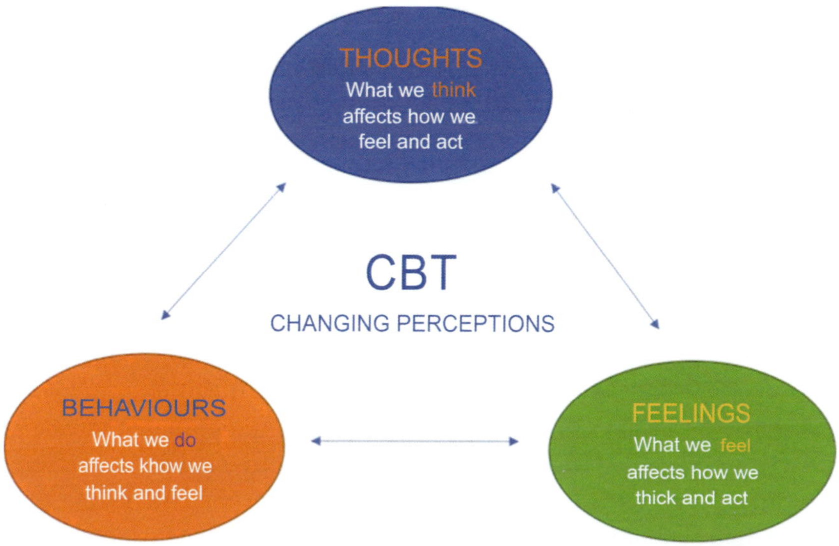

Figure 6.3. *CBT helps gain control, reduce symptoms, and develop effective strategies to deal with stress confidently.*

Source: CBT for Stress Management | KlearMinds (2024).

Behavioral approaches in CBT are also essential for managing stress. Activity scheduling and behavioral activation are two strategies that help combat the inactivity and withdrawal that frequently accompany stress and sadness in older persons by encouraging them to participate in fulfilling and meaningful activities. Mood and stress resilience are improved by increasing positive activity participation (Khoury et al., 2015).

Furthermore, CBT incorporates stress management techniques like relaxation training and problem-solving skills. Relaxation techniques, such as progressive muscle relaxation and deep breathing exercises, help reduce physiological arousal and promote a state of calm. Problem-solving training equips older adults with strategies to effectively address and manage daily stressors, fostering a sense of competence and control (Geiger et al., 2016).

The effectiveness of CBT for stress management in older individuals is supported by empirical research. According to Khoury et al. (2015), CBT dramatically lowers anxiety and depressive symptoms in senior citizens, enhancing their general well-being. Furthermore, CBT has been demonstrated to improve coping mechanisms and cognitive performance, which lessens the negative effects of stress (Geiger et al., 2016).

6.2.3. Social Engagement and Support

The management of stress in the older adult population is significantly influenced by social engagement and support networks, which offer practical and emotional advantages that reduce the negative consequences of stress. Numerous studies that show that social aspects have a favorable effect on older people's mental health and general well-being highlight how important they are (McCraty & Atkinson, 2012).

Firstly, social engagement fosters a sense of belonging and purpose, which is essential for psychological health. Participation in community activities, volunteer work, or social clubs helps older adults maintain active lifestyles and develop meaningful relationships, reducing feelings of loneliness and isolation that are prevalent in this age group. Loneliness is a significant stressor that exacerbates anxiety and depression; thus, increased social interaction can alleviate these negative emotions (McCraty & Atkinson, 2012). Engaging in social activities also provides cognitive stimulation, which can enhance mental acuity and resilience against stress-related cognitive decline.

Family, friends, and community groups make up support networks, which provide emotional support that reduces stress. By showing empathy, care, and encouragement, emotional support enables older persons to handle challenges more skillfully. The physiological effects of stress can be lessened by this encouraging relationship, which can lower cortisol levels (Glazer & Liu, 2017). Additionally, having a solid support system guarantees that senior citizens may get useful help when they need it, including assistance with daily duties or transportation, which lessens the stress that comes with handling these responsibilities on their own.

Moreover, social support is essential for improving coping strategies. Strong support networks increase the likelihood that people would use adaptive coping mechanisms like asking for help and solving problems instead of abusing drugs or other unhealthy habits (McCraty & Atkinson, 2012). Furthermore, social support can promote healthier lifestyle choices and enhance adherence to medical treatments, both of which further reduce stress and enhance health outcomes (Glazer & Liu, 2017). Empirical studies corroborate the protective effects of social engagement and support networks on stress. For instance, research has found that strong social relationships are associated with a 50% increased likelihood of survival, highlighting the critical role of social factors in longevity and stress management (McCraty & Atkinson, 2012).

6.2.4. Exercise and Physical Activity

Exercise and physical activity are essential for older persons to manage stress because they have significant psychological and physiological advantages. By regulating the HPA axis, which in turn lowers cortisol, the main stress hormone, regular exercise lowers stress levels (Sharon-David & Tenenbaum, 2017). This physiological action plays a critical role in reducing the negative effects that long-term stress has on the body (Figure 6.4).

Figure 6.4. *Exercise is crucial for older individuals to reduce stress.*

Source: Harvard Health (2020).

Exercise also stimulates the release of endorphins and other neurotransmitters such as serotonin and dopamine, which enhance mood and provide a sense of well-being. These biochemical changes alleviate symptoms of anxiety and depression, common among older adults, thus reducing overall stress levels (Sharon-David & Tenenbaum, 2017).

Furthermore, exercise enhances the quality of sleep, which is frequently compromised by stress. Improved sleep lowers stress levels even further by improving mood management and cognitive performance (McCraty & Atkinson, 2012). Regular exercise also fosters social contact because many physical activities, like walking clubs or group exercise courses, offer social chances that help people feel less alone and isolated, which can worsen stress-inducing feelings (Sharon-David & Tenenbaum, 2017).

Exercise helps older persons manage their stress, according to empirical data. In contrast to their sedentary counterparts, physically active older persons reported lower levels of perceived stress and higher overall mental health, according to Sharon-David & Tenenbaum's (2017) research. To sum up, physical activity and exercise provide older persons with a variety of stress management options by improving their mood, social relationships, physiological resilience, and quality of sleep.

6.3. COGNITIVE DECLINE AND STRESS

Chronic stress contributes to cognitive decline through multiple pathways that affect brain structure and function. Some mechanisms supported by scholarly sources include:

6.3.1. Hippocampal Atrophy

Prolonged stress raises cortisol levels, which can lead to hippocampal atrophy, a crucial part of the brain for memory and learning. Deficits in memory consolidation and retrieval result from this atrophy. According to research by Amarya et al. (2018), extended exposure to high cortisol levels is linked to a considerable reduction in hippocampus volume, which can lead to memory problems and cognitive impairments.

6.3.2. Oxidative Stress

Chronic stress increases the production of reactive oxygen species (ROS), leading to oxidative damage in the brain. This oxidative stress damages neurons and impairs synaptic function, contributing to cognitive decline. Oxidative stress is a significant factor in the pathogenesis of neurodegenerative diseases and cognitive decline (Birch-Machin & Bowman, 2016).

6.3.3. Impaired Neurogenesis

Chronic stress inhibits neurogenesis, particularly in the hippocampus, where new neurons are generated throughout life. This disruption hampers the brain's ability to form new memories and adapt to new information. Birch-Machin & Bowman (2016) demonstrated that stress-induced suppression of hippocampal neurogenesis is associated with cognitive impairments and increased vulnerability to mental disorders.

6.3.4. Prefrontal Cortex Dysfunction

Chronic stress has a negative impact on the prefrontal cortex, which controls executive processes like working memory, attention, and decision-making. Increased cortisol damages prefrontal neurons' structural integrity and decreases their connection. Chronic stress reduces prefrontal brain function, which results in deficiencies in cognitive control, attention, and emotional regulation, according to de Magalhães & Passos (2018).

6.3.5. Neuroinflammation

Interleukin-6 (IL-6) and tumor necrosis factor-alpha (TNF-α) are two pro-inflammatory cytokines that are elevated in the brain during prolonged stress. Neuroinflammation is associated with cognitive impairments and aggravates neuronal damage. According to Amarya et al. (2018), inflammation plays a part in mediating the development of neurodegenerative illnesses and stress-related cognitive deficits.

6.4. PHYSICAL HEALTH ISSUES AND STRESS IN AGING

6.4.1. Physical Health in Aging

The various physiological changes and health issues that people face as they age are all included in the concept of physical health in aging. A person's health and quality of life can be greatly impacted by the steady deterioration in physiological function which is a hallmark of aging (Cohen et al., 2016).

The loss of musculoskeletal health is one of the main issues associated with aging. The age-related loss of muscle mass and strength known as sarcopenia impairs mobility and raises the possibility of fractures and falls (Cohen et al., 2016). Moreover, osteoporosis, which is defined as a decrease in bone density, intensifies the vulnerability to fractures (Compston et al., 2019).

Cardiovascular health also deteriorates with age, contributing to a higher prevalence of conditions such as hypertension, coronary artery disease, and heart failure. These conditions are compounded by age-related changes in the heart and blood vessels, including arterial stiffening and reduced cardiac output (Yaribeygi et al., 2017). Similarly, respiratory function declines, with reduced lung capacity and efficiency, making older adults more vulnerable to respiratory infections and chronic obstructive pulmonary disease (COPD) (Yaribeygi et al., 2017).

Sensory impairments, such as diminished vision and hearing, are prevalent among older adults and can lead to difficulties in communication, social isolation, and a decreased quality of life (Kim et al., 2021). Moreover, the immune system's efficacy wanes with age, a phenomenon known as immunosenescence, which increases susceptibility to infections, malignancies, and autoimmune disorders (Yaribeygi et al., 2017).

Maintaining physical health in aging requires a multifaceted approach, including regular physical activity, a balanced diet, and preventive healthcare. Exercise has been shown to mitigate some of the adverse effects of aging by improving cardiovascular and musculoskeletal health (Sharon-David &

Tenenbaum, 2017). Nutritional interventions, particularly those rich in protein, calcium, and vitamins D and B12, are crucial for sustaining muscle and bone health (Bauer et al., 2013).

Addressing the complex interplay of physiological changes in aging is critical for enhancing health and well-being in older adults. A comprehensive strategy that includes lifestyle modifications and preventive care can significantly improve the quality of life for the aging population.

6.4.2. Stress in Aging

Stress as we age is a complex phenomenon that has important effects on our mental and physical well-being. As people get older, they come into contact with a variety of stresses that may worsen pre-existing health issues or cause the development of new ones. These pressures are frequently caused by psychological difficulties, social changes, and health-related problems.

Declining health is one of the main causes of stress as people age. According to Santini et al. (2020), chronic illnesses such as diabetes, arthritis, and cardiovascular disease are common in older persons and can cause ongoing discomfort, incapacity, and a decline in quality of life. Continuous medical treatment is necessary for the management of many chronic diseases, which can be extremely stressful, especially for people with little financial or healthcare resources.

Social changes also contribute to stress in aging. The loss of a spouse or close friends, retirement, and the consequent reduction in social interaction can lead to feelings of loneliness and social isolation (Golden et al., 2019). These social stressors are closely linked to mental health issues such as depression and anxiety, which are common in older populations (Santini et al., 2020).

Furthermore, cognitive decline is a critical factor in stress among the elderly. Mild cognitive impairment and dementia are associated with increased dependency and reduced autonomy, which can be profoundly distressing for both the individuals affected and their caregivers (Cummings et al., 2020). The fear of losing cognitive function and the stigma associated with cognitive decline further compound this stress.

Effective coping mechanisms and support systems are crucial in mitigating the impact of stress on aging. Social support, engagement in meaningful activities, and psychological interventions have been shown to alleviate stress and improve overall well-being in older adults (Park et al., 2020). Therefore, addressing stress in aging necessitates a comprehensive approach that encompasses medical, psychological, and social interventions.

6.4.3. Stress and Physical Health's Interconnection

Research on the connection between stress and physical health is essential, especially as people age. It has been demonstrated that long-term stress has a significant negative influence on physical health, aggravating pre-existing ailments and raising the risk of developing new ones. On the other hand, being in poor physical health can make stress levels higher, which can have a reciprocating effect that feeds the cycle of growing stress and declining health.

In a physiological sense, stress triggers the release of cortisol and other stress hormones by activating the hypothalamic-pituitary-adrenal (HPA) axis. Chronic inflammation is a known risk factor for many age-related diseases, such as diabetes, neurological disorders, and cardiovascular disease, and can be brought on by prolonged activation of the HPA axis (Birch-Machin & Bowman, 2016). Prolonged inflammation can further weaken the immune system, increasing the risk of infections and delaying recovery in older persons.

Moreover, stress can influence behaviors that negatively affect physical health. High-stress levels are associated with poor dietary choices, decreased physical activity, smoking, and alcohol consumption (Yaribeygi et al., 2017). These behaviors contribute to the development and progression of chronic illnesses. For instance, a study by Cohen et al. (2019) found that individuals with high-stress levels were at an increased risk of developing metabolic syndrome, a cluster of conditions that increase the risk of heart disease, stroke, and diabetes.

In turn, the presence of chronic physical illnesses can increase psychological stress. The burden of managing chronic diseases, coping with pain, and dealing with functional impairments can lead to heightened anxiety and depression (Yaribeygi et al., 2017). This psychological distress can exacerbate physical symptoms and hinder effective disease management, further perpetuating the cycle of stress and physical health decline.

Recognizing this relationship emphasizes how important it is to use integrated healthcare strategies that address mental and physical health, especially in older populations. Breaking this loop and enhancing general well-being can be especially successful with interventions that incorporate medical care with stress-reduction methods, such as mindfulness and cognitive-behavioral therapy (Khoury et al., 2015).

6.4.4. Methods for Promoting Ageing in a Healthy Way

Encouraging stress-free aging requires a multimodal strategy that incorporates social interaction, lifestyle changes, and all-encompassing healthcare tactics. Research indicates that three essential elements are mental health support, a balanced diet, and physical activity.

Regular physical activity is paramount in mitigating age-related decline and stress. Exercise has been shown to improve cardiovascular health, muscle strength, and flexibility, while also reducing anxiety and depression (Sharon-David & Tenenbaum, 2017). The American College of Sports Medicine recommends at least 150 minutes of moderate-intensity aerobic activity per week for older adults.

An important factor in healthy aging is nutrition. Diets like the Mediterranean diet, which are high in fruits, vegetables, whole grains, and lean proteins, have been associated with better mental health outcomes and a lower risk of chronic diseases (Davis et al., 2015). It's also critical to treat micronutrient deficiencies and drink enough water.

Support for mental health is essential for helping older people manage their stress. Anxiety and depressive symptoms have been successfully reduced by cognitive-behavioral therapy (CBT) and mindfulness-based stress reduction (MBSR) (Santini et al., 2020). Strong support systems and social interaction can act as stress reducers by fostering a feeling of purpose and belonging. Research shows that social isolation poses a serious risk to older persons' mental health and increases their mortality (Santini et al., 2020).

Access to comprehensive healthcare services, including regular check-ups and preventive care, is vital. Multidisciplinary care teams can address the complex health needs of older adults, ensuring timely management of chronic conditions and reducing stress related to health concerns.

6.5. STRESS AND SOCIAL ISOLATION

The absence of social interactions, contacts, and ties with family, friends, and neighbors on a personal level as well as with society at large on a broader level is referred to as social isolation. It is a condition in which a person interacts with people infrequently and doesn't have deep relationships that offer emotional support and a feeling of community. Regardless of the quantity of social interaction, loneliness is a subjective experience of being alone; social isolation is not the same as loneliness. While social isolation is an objective state of little social engagement, loneliness can nonetheless exist in social situations (Park et al., 2020).

6.5.1. Causes of Social Isolation

6.5.1.1. Demographic Shifts

1. **Ageing Population:** As people get older, they frequently lose spouses, parents, and friends, which causes their social networks to get smaller. Furthermore, living alone is increasingly common among older persons, especially in cases of widowhood or separation.

2. **Family Dynamics:** According to Leigh-Hunt et al. (2017), smaller families and greater geographic dispersion of family members can lead to a decline in familial support and an increase in isolation.

6.5.1.2. Health-Related Issues

1. **Chronic Illness and Disability:** Isolation can result from physical health issues, long-term illnesses, and impairments that restrict movement and social interaction.
2. **Mental Health Conditions:** According to Leigh-Hunt et al. (2017), mental health conditions including anxiety and depression can cause social disengagement and a sense of loneliness.

6.5.1.3. Cultural and Societal Aspects

1. **Technological Advancements:** Although technology can bring people together, overuse of digital devices can obviate in-person connections, paradoxically increasing feelings of isolation (Santini et al., 2020).
2. **Cultural Norms:** Societal expectations and cultural norms around aging and socialization can influence the degree to which individuals remain socially engaged. In some cultures, older adults may be more likely to be marginalized and isolated (Kim et al., 2021).

6.5.1.4. Environmental Factors

1. **Geographic Location:** Living in rural or remote areas can limit access to social networks and community resources, increasing the risk of isolation (Kim et al., 2021).
2. **Safety and Accessibility:** Neighborhood safety and the accessibility of public spaces and transportation play a critical role in enabling social interactions. Unsafe or inaccessible environments can deter individuals from participating in social activities (Park et al., 2020).

Understanding the multifaceted causes of social isolation is essential for developing targeted interventions to mitigate its impact on individuals, particularly those in vulnerable populations. By addressing the underlying factors contributing to social isolation, it is possible to foster greater social inclusion and improve overall well-being.

6.5.2. The Impact of Social Isolation on Health

Social isolation has significant negative effects on one's physical, mental, and cognitive well-being. Academic studies demonstrate that social isolation raises the risk of several negative health outcomes by a significant amount.

6.5.2.1. Implications for Mental Health

There is a clear correlation between declining mental health and social isolation. According to research by Goll et al. (2015), mental illnesses like depression and anxiety can arise and worsen because loneliness, a subjective indicator of social isolation, plays a role in these processes. Feelings of loneliness can grow in a lack of social engagement and support, creating a vicious cycle that deteriorates mental health. Goll et al. (2015) also showed that those who are socially isolated experience higher levels of perceived stress and fewer coping mechanisms, which exacerbates mental health issues.

6.5.2.2. Physical Health Consequences

The physical health ramifications of social isolation are equally significant. Studies have shown that socially isolated individuals have a heightened risk of chronic illnesses such as cardiovascular disease, hypertension, and type 2 diabetes. Holt-Lunstad, Smith, & Layton (2010) conducted a meta-analysis revealing that social isolation increases the likelihood of mortality by 29%, a statistic that underscores its severe impact. Socially isolated individuals often exhibit higher levels of inflammatory markers, such as C-reactive protein, which are predictors of cardiovascular problems (Steptoe et al., 2004). Furthermore, social isolation can lead to poor health behaviors, such as physical inactivity, poor diet, and substance abuse, which exacerbate chronic conditions (Shankar et al., 2011).

6.5.2.3. Cognitive Decline

Cognitive health is also adversely affected by social isolation. Khoury et al. (2015) found that social isolation is associated with an accelerated rate of cognitive decline in older adults. The lack of cognitive stimulation and social engagement contributes to diminished cognitive function and increases the risk of developing dementia. A study by James et al. (2011) concluded that socially isolated individuals had significantly higher rates of cognitive impairment, suggesting that social interaction plays a crucial role in maintaining cognitive health.

6.5.2.4. Mechanisms Underlying the Impact on Health

The harmful consequences of social isolation on one's health are explained by several processes. According to Yaribeygi et al. (2017), social isolation can biologically cause chronic stress responses that prolong the increase of stress hormones like cortisol, which can be harmful to many body systems. Social isolation can psychologically cause feelings of hopelessness and unworthiness, which can have a detrimental effect on mental health and raise health risks. In terms of behavior, the lack of social support frequently leads to a disregard for health-promoting practices and a rise in dangerous behavior.

6.5.2.5. Mitigation Strategies

Addressing the health impacts of social isolation requires multi-faceted approaches. Social interventions, such as community engagement programs, are crucial for providing social support and enhancing connectedness. Healthcare providers should be vigilant in screening for social isolation and incorporating social health into their assessments and care plans. Technological solutions, such as telehealth and virtual social networks, can also help mitigate the effects of social isolation, particularly in hard-to-reach populations.

6.5.3. The Relationship Between Social Isolation and Stress

A lack of social interaction or involvement is known as social isolation, and it is becoming more widely acknowledged that social isolation poses a serious risk for stress in a variety of demographic contexts. The significant influence of social connectivity on both psychological and physical well-being is highlighted by this link.

6.5.3.1. Emotional and Psychological Aspects

Due to a lack of supportive social interactions, social isolation can result in increased psychological stress. Perceived social isolation, according to Park et al. (2020) research, causes a stress response in the brain similar to that of physical pain, inducing the production of stress hormones like cortisol and activating neural circuits linked to threat perception. These physiological reactions can worsen emotions of loneliness and mental anguish as well as prolong a stressful cycle (Park et al., 2020).

Furthermore, the psychological toll of social isolation manifests in increased susceptibility to mood disorders such as depression and anxiety. Studies by Holt-Lunstad et al. (2015) have shown that loneliness and social isolation are associated with a higher risk of developing depressive symptoms, independent of other factors. Individuals experiencing social isolation may also perceive their environments as more threatening, further heightening stress levels and negatively impacting their overall mental health (Park et al., 2020).

6.5.3.2. Physiological Implications

Beyond psychological effects, social isolation influences physiological health through mechanisms linked to chronic stress. Prolonged activation of stress pathways, as observed in socially isolated individuals, contributes to systemic inflammation, compromised immune function, and cardiovascular dysregulation (Amarya et al., 2018). This physiological dysregulation increases vulnerability to chronic diseases such as hypertension, diabetes, and cardiovascular disorders, all of which are exacerbated by stress-related mechanisms.

Moreover, the impact of social isolation on physiological health extends to cognitive functioning. Older adults experiencing social isolation exhibit accelerated cognitive decline, possibly mediated by stress-induced changes in brain structure and function (Amarya et al., 2018). Research suggests that social engagement serves as a protective factor against cognitive impairment, highlighting the intricate interplay between social connectedness, stress modulation, and cognitive resilience (Chu et al., 2024). These dynamics inform targeted strategies that foster inclusive and supportive social environments conducive to optimal health outcomes.

6.5.4. How to Deal with Social Isolation and Stress?

To effectively address social isolation and stress in people, particularly in vulnerable populations like older seniors, an integrated approach that incorporates social, psychological, and health-related techniques is needed. Scholarly literature and research offer valuable perspectives on successful strategies designed to help ease these difficulties.

First, developing deep social ties is a crucial part of combating social isolation. A 2015 study by Goll et al. found that social ties had a major impact on health outcomes, with social isolation and loneliness being associated with a higher risk of death that is similar to that of obesity and smoking. According to Goll et al. (2015), community-based programs, support groups, and intergenerational activities should be the main focus of interventions aimed at improving social networks. One way to reduce loneliness and improve emotional well-being is through regular social interaction, which is facilitated by programs like senior centers or neighborhood get-togethers.

Psychologically, managing stress associated with social isolation requires coping mechanisms that promote resilience and emotional regulation. Kim et al. (2021) transactional model of stress and coping emphasizes the importance of cognitive appraisal and adaptive coping strategies in stress management. Cognitive-behavioral interventions, as outlined by Hawkley & Cacioppo

(2018), focus on altering maladaptive thought patterns and enhancing problem-solving skills to mitigate stressors associated with social isolation. Health-related approaches involve integrating physical and mental health care to address both the psychological and physiological impacts of isolation and stress. For example, healthcare professionals can incorporate routine screenings for depression and anxiety among isolated individuals, as suggested by Masi et al. (2017), to provide timely interventions. Moreover, promoting healthy lifestyles through exercise programs, nutritional counseling, and mindfulness-based stress reduction techniques (Ozbay et al., 2007) can bolster resilience and improve overall well-being.

Policy plays a pivotal role in supporting these interventions. Effective policies should prioritize funding for community-based social services, accessible transportation options, and age-friendly infrastructure that facilitates social participation (Cohen-Mansfield et al., 2016). Advocacy efforts aimed at raising awareness about the detrimental effects of social isolation and stress can also influence policy decisions and resource allocation.

6.6. FINDING PURPOSE AND MEANING IN LATER LIFE

When you are under stress, one way to find meaning and purpose is to figure out the reason why are having the problems you face. It frequently entails realizing how such difficulties fit in with your beliefs, objectives, and ultimate life purpose. Linking what you're going through, and your priorities can provide you resilience, direction, and the drive to get through stressful situations. To establish meaning and purpose in your life, this process frequently entails reflection on personal values, introspection, and figuring out how to match your daily actions with those principles. Some of the ways an individual can find purpose and meaning in later life are discussed in subsections.

6.6.1. Social Connection and Relationship

In later life, social contacts and connections are essential for finding meaning and purpose, especially when it comes to reducing stress. Significant life transitions associated with aging, such as retirement, grief, and deteriorating health, can exacerbate stress and leave one feeling aimless. However social relationships—which offer possibilities for meaningful participation, emotional support, and a sense of belonging—can mitigate these difficulties (Figure 6.5).

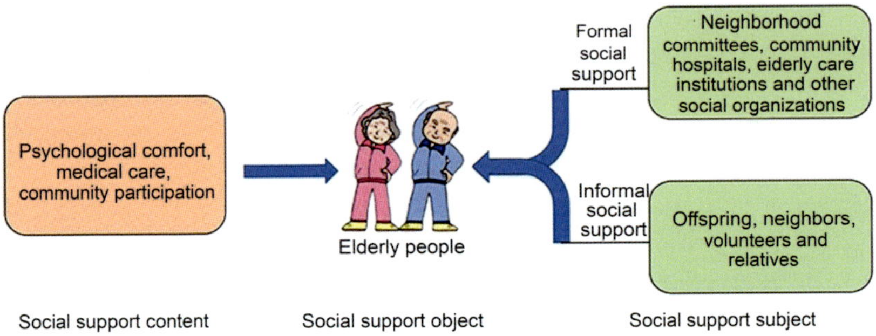

Figure 6.5. *Social support on the mental health of elderly individuals.*

Source: Zhang & Dong (2023).

The role of social connections in later life is extensively documented in scholarly literature. According to a study by Boszormenyi-Nagy (2014), social relationships are fundamental to psychological well-being and stress reduction among older adults. The authors argue that social networks offer emotional support, practical assistance, and opportunities for meaningful social roles, which are essential for maintaining a sense of purpose and reducing feelings of isolation and stress.

Emotional support from social connections helps older adults cope with stress by providing a buffer against the negative effects of life changes and health issues. For instance, having someone to talk to during times of stress can alleviate feelings of loneliness and anxiety. The presence of close, supportive relationships has been linked to lower levels of cortisol, a stress hormone, thereby contributing to better physical and mental health (Santini et al., 2020).

Social connections also provide a feeling of belonging, which is essential for discovering meaning and purpose. Since humans are social creatures by nature, having a sense of belonging to a community or social group can strengthen one's sense of self-worth and identity. The need to belong is a basic human motivator, according to Randall & Bodenmann's (2017) "belongingness hypothesis," and satisfying this need is linked to several advantageous outcomes, such as lowered stress and increased life satisfaction.

Engagement in social activities and roles also provides older adults with opportunities to find meaning. Participating in group activities, volunteering, or taking on caregiving roles allows older individuals to contribute to others' well-being, fostering a sense of purpose. Volunteering and other forms of social participation are associated with greater life satisfaction and lower levels of depression among older adults. These activities not only keep older adults physically and mentally active but also provide a sense of achievement and purpose. Furthermore, intergenerational relationships can play a significant role in finding purpose in later life. Engaging with younger generations, whether through family interactions or community programs, allows older adults to pass on knowledge, skills, and traditions, which can be highly meaningful. Studies have shown that such interactions can enhance the well-being of older adults by affirming their sense of legacy and continued relevance in society (Kim et al., 2021).

6.6.2. Participation and Contribution

Contribution and engagement are essential for determining meaning and purpose in later life, and they have a big impact on older persons' stress levels. People experience a variety of pressures as they get older, such as health issues, social isolation, and life transitions. These pressures have the power to erode one's sense of direction and bring on depressing and gloomy thoughts. Active participation and the capacity to make a significant contribution to society, on the other hand, might mitigate these impacts and enhance psychological well-being while lowering stress.

Engagement in meaningful activities provides structure and a sense of routine, which are essential for maintaining mental health in older adults. According to a study by Kelly et al. (2019), participation in purposeful activities such as volunteering or part-time work is associated with lower levels of depressive symptoms and higher levels of life satisfaction among older adults. Engaging in such activities helps individuals maintain a sense of identity and self-worth, which are crucial components of mental health and stress resilience.

In addition, the sense of contribution that comes from involvement cultivates a sense of worth and necessity, which can operate as a powerful protective barrier against the adverse consequences of stress associated with aging. According to Santini et al. (2020), who highlight the value of productive engagement in their model of successful aging, engaging in worthwhile activities can enhance one's sense of significance and accomplishment. This sense of accomplishment can be especially helpful in reducing feelings of social isolation and redundancy, which are frequent sources of stress at a later age.

Social engagement also plays a critical role in reducing stress and enhancing a sense of purpose. Active participation in community and social groups provides opportunities for social interaction, reducing feelings of loneliness and isolation. Santini et al. (2020) found that social relationships are strongly linked to mental health outcomes, with greater social integration associated with lower levels of stress and better overall well-being. Engaging in group activities can thus foster social bonds and provide emotional support, further enhancing an older adult's sense of purpose and meaning.

Participating in cognitive activities and lifelong learning can also assist preserve cognitive function and postpone cognitive decline, which is a major cause of stress for a lot of older persons. Engaging in cognitive activities like reading, going to classes, and taking part in thought-provoking conversations is linked to improved cognitive health and lower levels of stress.

Therefore, making a contribution and being involved is essential to discovering meaning and purpose in later life, which has a direct impact on older individuals' stress levels. Through the development of a sense of self-worth, routine, social interaction, and cognitive engagement, these activities assist in reducing the stressors linked to aging and advance psychological health.

6.6.3. Cultivating Hobbies and Activities of Interest to an Individual

Developing interests and hobbies is essential to discovering meaning and purpose in later life, especially when it comes to stress reduction. Stress can be reduced, and psychological well-being can be considerably increased by partaking in activities that bring a sense of success. Emotional and mental health depends on people continuing to actively pursue their interests as they get older since this keeps them feeling like they have a purpose and identity (Goll et al., 2015).

According to the activity theory of aging, staying active and involved is crucial for life satisfaction and overall well-being in older adults (Boszormenyi-Nagy, 2014). Hobbies and interests offer a structured and enjoyable way to remain engaged, countering the potential isolation and inactivity that can accompany retirement and other age-related transitions. This active engagement can buffer against the stressors associated with aging, such as loss of professional identity and reduced social interactions (Boszormenyi-Nagy, 2014).

Additionally, hobbies frequently present chances for socialization, which is essential for lowering stress and fostering mental wellness in later life. Participating in community gardening or joining a book club are two examples of group activities that promote social connections and a sense of belonging. According to Goll et al. (2015), these social networks can offer practical help, emotional support, and a stress-reduction mechanism.

The concept of flow, as introduced by Csikszentmihalyi (1990), further underscores the importance of hobbies in finding meaning. Flow experiences, characterized by deep immersion and enjoyment in an activity, contribute to intrinsic motivation and a sense of fulfillment. For older adults, engaging in flow-inducing activities can enhance the quality of life and provide a profound sense of purpose, counteracting feelings of emptiness and stress (Nakamura & Csikszentmihalyi, 2002).

6.7. RESOURCES FOR OLDER ADULTS DEALING WITH STRESS

Unique stressors associated with aging, health problems, and social isolation are frequently experienced by older persons. Tools designed specifically for this group of people can play a significant role in effective stress management. With the help of these resources, people can improve their general and mental health on their initiative. This talk covers several important self-help techniques that academic research has validated.

6.7.1. Online Sources

6.7.1.1. Mindfulness and Meditation Apps

The benefits of mindfulness and meditation in lowering stress have been thoroughly researched. Seniors can benefit especially from guided meditation sessions offered by apps like Headspace and Calm. Studies reveal that engaging in mindfulness techniques can considerably reduce stress, promote emotional regulation, and improve the quality of life for senior citizens (Black et al., 2015). These apps frequently come with tailored programs that deal with certain issues that older persons frequently face, like anxiety, insomnia, and chronic pain.

6.7.1.2. Reading Materials

Books and articles focused on stress management provide valuable insights and practical advice. Cognitive-behavioral techniques (CBT) are frequently discussed in these materials. CBT-based self-help books can effectively reduce symptoms of anxiety and depression, common in stressed older adults. Books like "The Relaxation and Stress Reduction Workbook" by Davis, Eshelman, and McKay offer a variety of techniques that can be easily implemented into daily routines (Khoury et al., 2015).

6.7.1.3. Online Support Groups

Many online support groups are available on the internet. These groups give a shared experience and a sense of community, which may be very helpful for older individuals who are stressed. You can find these groups on social media sites like Facebook and specialized websites like AgingCare.com. According to research by White & Dorman (2021), participating in online support groups can help people feel less alone and provide them with emotional support, which can help them cope with stress.

6.7.1.4. Exercise Programs

Physical activity is a well-documented stress reliever. Online exercise programs, specifically designed for older adults, can be accessed through platforms like YouTube or subscription-based services. Tai Chi and yoga are particularly effective, as they combine physical movement with mindfulness and breathing techniques. Regular participation in Tai Chi significantly reduces stress and improves psychological well-being in older adults (Geiger et al., 2016).

6.7.1.5. Digital Journals

Creating a journal is a therapeutic practice that aids in mental and emotional processing. For older folks, digital journals—which can be accessed through applications like Day One—provide a practical means of participating in reflective writing. According to a 2017 study by Ullrich and Lutgendorf, writing expressively can improve mental health and reduce stress. The inclusion of prompts and reminders in these notebooks helps users stick to a regular writing schedule.

6.7.1.6. Educational Websites and Videos

These resources include articles, videos, and interactive content that can educate older adults about the causes of stress and effective coping strategies. The accessibility and reliability of these sources make them invaluable tools for self-education.

6.7.1.7. Agencies for Mental Health

Mental health organizations are essential in helping senior citizens manage their stress because they provide a comprehensive strategy that takes into account the particular difficulties that this group faces. These organizations give older people access to professional counseling and therapy, which is crucial for assisting them in processing and coping with age-related stressors like retirement, health problems, and bereavement. Reframing negative thought patterns and developing effective coping mechanisms are common goals of evidence-based psychotherapies, such as mindfulness-based stress reduction (MBSR) and cognitive behavioral therapy (CBT) for older individuals (Gonçalves & Byrne, 2013).

Furthermore, support groups designed exclusively for senior citizens are provided by mental health organizations. These groups give a forum for exchanging experiences and offering assistance to one another, which helps lessen stress and feelings of loneliness (Williams et al., 2015). The sense of community and belonging that these groups promote is essential to emotional health. Agencies also put a lot of effort into psychoeducation, which gives older persons the information they need to understand stress and how it affects their health and take proactive steps to control it. They frequently work in tandem with medical professionals to guarantee a comprehensive strategy that integrates mental and physical health care. This is especially advantageous because stress can aggravate chronic health disorders that are common in the elderly (Chu et al., 2024).

Moreover, mental health agencies advocate for and provide access to community resources, such as social services and recreational activities, which enhance the overall quality of life and reduce stress. By addressing both the psychological and social dimensions of stress, mental health agencies offer comprehensive support that is crucial for the well-being of older adults.

6.7.2. Community Wellness Centers for Older Adults

Community centers that provide wellness programs are essential for helping senior citizens manage their stress. These facilities offer a variety of programs designed to meet the unique requirements of senior citizens, offering a comprehensive approach to stress management. Seniors' psychological and physical well-being is greatly enhanced by participating in community-based wellness activities, according to a study by Aung & Friedman (2019).

First and foremost, stress reduction is greatly aided by physical exercises like tai chi, yoga, and low-impact aerobics that are offered at community centers. These activities enhance physical health, which has a direct bearing on mental wellness. Studies reveal that engaging in consistent physical activity lowers cortisol levels, a hormone linked to stress, and increases endorphin production, resulting in elevated mood and decreased anxiety (Park et al., 2020).

Secondly, community centers often provide social engagement opportunities through group activities, which are essential for emotional support. Social isolation is a significant stressor among older adults, and participation in community activities fosters a sense of belonging and reduces feelings of loneliness. A study by Cacioppo & Cacioppo (2014) highlights that social interaction can lower stress levels by providing emotional support and enhancing coping mechanisms.

Furthermore, educational classes and support groups are essential parts of programs offered by community centers. These seminars provide helpful information on coping mechanisms, healthy lifestyle options, and stress management practices. They also offer a forum for experience sharing, which has the potential to be healing. Perceived social support from these contacts functions as a buffer against the negative consequences of stress, according to Goll et al. (2015).

Moreover, it has been demonstrated that mindfulness and meditation classes—which are becoming more and more accessible at community centers—significantly lower stress. Engaging in mindfulness activities can improve emotional control and induce peace. Programs for mindfulness-based stress reduction (MBSR) are successful in lowering symptoms of stress, anxiety, and depression, according to Geiger et al. (2016).

For elderly people, community centers that offer wellness programs offer a comprehensive approach to stress management. These centers address multiple aspects of well-being, including physical health, social relationships, education, and mindfulness programs, which help older persons efficiently handle stress.

CHAPTER 7

STRESS AND CHILDREN

Contents

The phenomenon of stress among children has garnered significant attention within academic and clinical research due to its profound implications for child development and well-being. Stress in children can originate from a multitude of sources, including academic pressures, family dynamics, peer relationships, and environmental factors American Psychological Association (APA), 2020) (Figure 7.1).

Figure 7.1. *Understanding stress in children is crucial for interventions.*

Source: Kiddipedia (2024).

Studies reveal that children undergo stress reactions that might affect their physical and mental well-being, just like adults do (Masarik & Conger, 2017). Anxiety, depression, behavioral issues, and physical symptoms like headaches or stomachaches are just a few ways that stress can show up (APA, 2020). Chronic stress can affect attention, memory, and executive function, according to research on the effects of stress on children's cognitive development and academic achievement (APA, 2020).

Furthermore, the home environment has played an essential role in reducing or increasing children's stress. Children's perceptions of and responses to stress are significantly influenced by parenting practices, family dynamics, and socioeconomic levels (McLaughlin et al., 2015). Resilience in children exposed to stressors can be fostered by protective factors such as supportive family ties and efficient communication. Another important setting where children experience stress is school. A child's stress levels can be influenced by extracurricular activities, interpersonal relationships, and academic obligations. According to McLaughlin et al. (2015), schools are essential in detecting children who are under stress and offering them resources like counseling services, stress management courses, and a safe environment.

The introduction of social media and technology in recent years has given children's stressors additional dimensions. According to Masarik & Conger (2017), three new stressors have evolved in the modern era: cyberbullying, exposure to improper content, and the obligation to maintain a digital persona. These stressors necessitate new approaches to management and intervention.

Given the multifaceted nature of stress in children, interdisciplinary approaches involving psychology, education, and social work are essential for developing comprehensive strategies to support children's mental health. Continued research and the implementation of evidence-based practices are imperative to mitigate the adverse effects of stress and promote healthy development in children (APA, 2020).

7.1. EFFECTS OF STRESS ON CHILDREN'S HEALTH

Chronic stress can profoundly impact children's health, leading to various detrimental effects. The commonest effects of stress on children's health are as follows:

7.1.1. Physiological Health Problems

Chronic stress in children has significant physiological implications, affecting their overall health and development. Persistent exposure to stress activates the hypothalamic-pituitary-adrenal (HPA) axis, leading to prolonged secretion of cortisol and other stress hormones. This continuous hormonal imbalance can disrupt various bodily systems, contributing to several health issues (APA, 2020).

An important physiological consequence is a higher risk of obesity. Because chronic stress raises cortisol levels, it alters metabolism and encourages the storage of fat, especially visceral fat. Obesity and related health issues, including type 2 diabetes and cardiovascular illnesses, can result from this metabolic disruption (McLaughlin et al., 2015).

Furthermore, long-term stress can weaken children's immune systems, making them more vulnerable to infections and diseases. Prolonged stress reduces immune function and the body's capacity to properly fight infections. According to McLaughlin et al. (2015), immunosuppression makes common illnesses like the flu and colds more common and severe.

Chronic stress also affects the cardiovascular system by sustaining a heightened state of arousal. Increased blood pressure and heart rate brought on by elevated stress hormones can damage blood vessels over time and lead to the development of hypertension and other cardiovascular diseases (APA, 2020).

Lastly, digestive issues may result from chronic stress. Irritable bowel syndrome (IBS) and other digestive diseases can be brought on by stress-induced alterations in gut motility and secretion as well as an imbalance in the gut microbiota (Masarik & Conger, 2017).

7.1.2. Psychological Distress

Chronic stress in children is a significant contributor to psychological distress, leading to a range of mental health issues. When children are exposed to prolonged stressors, their bodies produce elevated levels of stress hormones such as cortisol. Persistent activation of the stress response system can disrupt the normal functioning of the brain, particularly areas involved in emotion regulation and cognitive processing (Barrero-Castillero et al., 2019).

The emergence of anxiety and depression disorders is one of the main effects of chronic stress. Children who are under constant stress may show signs of increased anxiety, such as excessive concern, fear, and avoidance behaviors. Furthermore, depressive symptoms such as ongoing sadness, a lack of interest in activities, and feelings of hopelessness might result from long-term stress (Barrero-Castillero et al., 2019).

Chronic stress may also inhibit a child's capacity to control their emotions, which can lead to increased irritability, anger, and behavioral issues. According to McLaughlin et al. (2015), emotional dysregulation might worsen psychological suffering and make it harder to interact socially and function well in school.

In severe cases, chronic stress can contribute to the development of post-traumatic stress disorder (PTSD) in children, particularly those exposed to traumatic events. PTSD is characterized by intrusive memories, nightmares, and severe anxiety, which significantly impair daily functioning (Barrero-Castillero et al., 2019).

7.1.3. Behavioral Problems

Children who experience chronic stress frequently exhibit severe behavioral problems that have a major negative influence on their socialization and development. Long-term exposure to stressors triggers the HPA axis, which increases the release of stress hormones like cortisol. The prefrontal cortex and amygdala are two brain regions that are involved in regulating behavior that may be impacted by a hormonal imbalance (Masarik & Conger, 2017). Increased aggression, hyperactivity, and oppositional behaviors are some of the behavioral signs of chronic stress in children. These externalizing behaviors, which can result in confrontations with peers and authority people, are frequently coping methods that kids acquire in response to stress (McLaughlin et al., 2015). In addition, a child's behavioral profile may become more complicated due to internalizing behaviors like depression, anxiety, and withdrawal brought on by prolonged stress (Masarik & Conger, 2017).

The impact of chronic stress on behavior is particularly evident in children living in adverse environments, such as those experiencing poverty or family dysfunction. These children are more likely to exhibit behavioral problems due to continuous exposure to stressors and the lack of supportive resources to mitigate the effects of stress (Masarik & Conger, 2017). Addressing chronic stress through early intervention and supportive environments is crucial to prevent and manage these behavioral issues.

7.1.4. Impaired Cognition

Children under chronic stress exhibit significantly decreased cognitive function, which has long-term developmental effects. Chronic stress causes the HPA axis to remain activated, which raises cortisol levels. Cortisol is a stress hormone that negatively impacts the structure and function of the brain. Chronic stress is specifically linked to decreased volume in brain areas that are vital for cognitive skills, such as the hippocampus and prefrontal cortex, which are crucial for memory and executive abilities, respectively (Masarik & Conger, 2017).

Children exposed to chronic stress exhibit deficits in various cognitive domains, including attention, working memory, and cognitive flexibility. These cognitive impairments can hinder academic performance and overall intellectual development. For instance, Masarik & Conger (2017) found that children from low socioeconomic backgrounds, who often experience chronic stress, show significant working memory deficits compared to their less-stressed peers.

Chronic stress can also cause long-lasting impacts on cognitive function by interfering with the neuronal circuitry of the brain, especially at critical stages of brain development (McEwen, 2017). This disruption can exacerbate the cognitive difficulties brought on by ongoing stress by impairing the child's capacity to process information, solve problems, and successfully regulate emotions.

7.2. SIGNS OF STRESS IN CHILDREN

Chronic stress in children can manifest through various signs, impacting their mental, emotional, and physical well-being. The most common signs of stress in children are discussed in subsections (Figure 7.2).

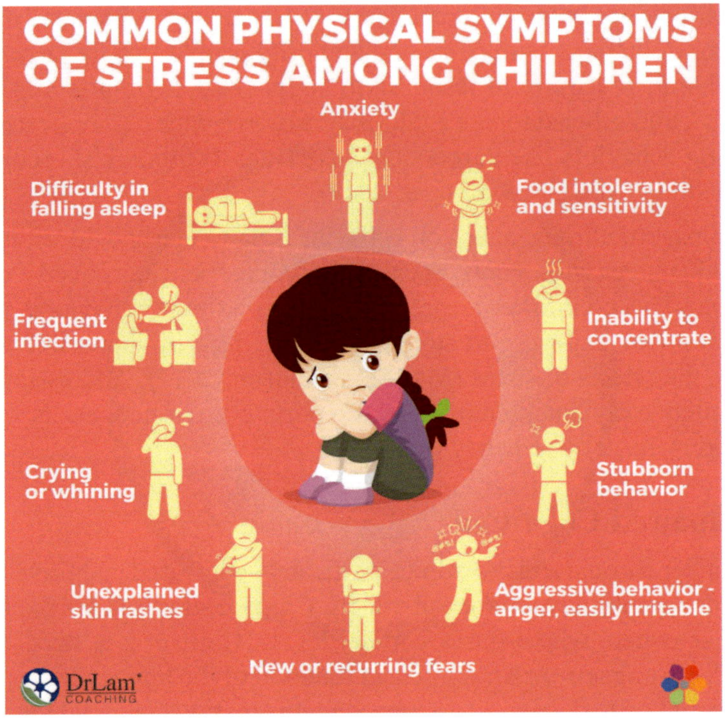

Figure 7.2. *How does stress affect children?*

Source: Robby (2021).

7.2.1. Behavioral Changes

Behavioral changes are a significant indicator of stress in children, reflecting their struggle to cope with ongoing stressors. These changes can manifest as increased irritability, aggression, and defiance, often disrupting their social and academic interactions. Chronic stress activates the HPA axis, leading to elevated cortisol levels, which can impair brain areas responsible for behavior regulation (APA, 2020).

Regressive habits, such as bedwetting or thumb-sucking, are attempts by children who are under constant stress to find security and comfort. These behavioral problems may result from internal stress responses like anxiety and dissatisfaction as well as external stressors like family disputes or academic expectations (Wunsch et al., 2019).

Furthermore, children who experience chronic stress may become withdrawn and lose interest in things they used to enjoy because they feel overwhelmed or unable to handle their stress. According to Barrero-Castillero et al. (2019), behavioral changes and emotional distress are frequently linked, emphasizing the need for early intervention and supportive environments to lessen the negative impacts of stress on children's development.

7.2.2. Emotional Dysregulation

Emotional dysregulation in children, often a sign of chronic stress, refers to the inability to manage emotional responses appropriately and consistently. Children experiencing chronic stress can exhibit heightened emotional reactivity, frequent mood swings, and difficulty calming down after being upset. This dysregulation is a consequence of the prolonged activation of the HPA axis, which results in elevated cortisol levels and disrupts the functioning of brain regions involved in emotion regulation, such as the prefrontal cortex and amygdala (McLaughlin et al., 2015).

Children who are emotionally dysregulated frequently struggle to adjust to changing circumstances and recover from poor emotional states. They may exhibit increased anxiety, impatience, and melancholy. According to Barrero-Castillero et al. (2019), these symptoms have the potential to worsen stress, which can set off a vicious cycle that worsens emotional and psychological health. Additionally, emotional dysregulation brought on by prolonged stress is linked to an increased risk of developing psychopathological diseases, such as anxiety and depression, underscoring the significance of early intervention and support in reducing these consequences (McLaughlin et al., 2015).

7.2.3. Physical Symptoms

Chronic stress in children often manifests through various physical symptoms, which serve as indicators of underlying psychological distress. These physical symptoms include headaches, stomachaches, muscle tension, and fatigue, frequently arising without any apparent medical cause. The continuous activation of the HPA axis due to chronic stress leads to elevated levels of cortisol and other stress hormones, which can disrupt normal bodily functions and contribute to these somatic complaints (Masarik & Conger, 2017).

Children who are under stress are more likely to experience headaches and stomachaches because this dysregulation of the autonomic nervous system, which regulates body functions like blood flow and digestion, occurs. Migraines and gastrointestinal problems may arise from this imbalance (APA, 2020). Chronic stress can also impair immunity, increasing a child's vulnerability to infections and diseases and adding to their physical discomfort (Masarik & Conger, 2017).

These physical symptoms can seriously affect a kid's quality of life and ability to function on a daily basis since they not only show that stress is there but also make the youngster feel more stressed overall. To lessen these impacts and promote the child's physical and mental health, early detection and intervention are essential (APA, 2020).

7.2.4. Sleep Disturbances

Sleep disturbances in children are a common sign of stress, reflecting the impact of psychological distress on physiological functions. Chronic stress activates the HPA axis, leading to elevated cortisol levels, which can disrupt sleep patterns. This disruption can manifest as difficulties falling asleep, frequent awakenings, nightmares, or restless sleep (Barrero-Castillero et al., 2019).

Sleep disruptions are more common among children exposed to stressful situations, such as those with family conflict or pressure to perform well academically, according to research. These sleep problems can intensify stress, resulting in a vicious cycle that has an impact on one's physical and emotional well-being (APA, 2020). Additionally, according to Barrero-Castillero et al. (2019), sleep disorders are associated with a higher risk of anxiety and sadness, emotional control issues, and diminished cognitive abilities.

Improving children's general well-being requires addressing sleep disruptions through stress-reduction strategies including cognitive-behavioral therapy, mindfulness exercises, and setting up a stable, supportive environment (Wunsch et al., 2019).

7.2.5. Academic Difficulties

Academic difficulties are a common manifestation of stress in children, significantly impacting their educational progress and overall well-being. Chronic stress can impair cognitive functions such as attention, memory, and problem-solving skills, which are critical for academic success. This impairment is primarily due to the prolonged activation of the HPA axis, leading to elevated levels of cortisol that disrupt neural functioning (Barrero-Castillero et al., 2019).

Stress related to school can cause children to lose focus, which can result in unfinished or substandard assignments. According to Masarik & Conger (2017), they frequently have a poorer capacity for memory, which has a negative impact on their academic performance and grades. In addition, the stress of having to perform well academically can intensify depressive and anxious feelings, leading to a vicious cycle in which stress impedes academic achievement, which in turn causes more stress (Wunsch et al., 2019).

According to Barrero-Castillero et al. (2019), children's stress levels can be greatly increased by the school environment itself, which includes elements like high expectations, standardized testing, and competitive environments. In order to help kids manage the demands of school and achieve better academic results, supportive educational methods and stress management initiatives must be implemented.

7.3. PARENTING STRATEGIES FOR MANAGING STRESS IN CHILDREN

Parents can employ several strategies to manage chronic stress in their children effectively. The strategies provided below are grounded in evidence-based practices and academic research:

7.3.1. Establishing a Supportive Environment

Establishing a supportive environment by parents plays a crucial role in managing chronic stress in children. A stable and nurturing home environment provides children with a sense of security and predictability, which is essential for their emotional well-being. When children feel safe and supported, they are better equipped to handle stress and develop resilience (Wunsch et al., 2019).

Regular routines, honest communication, and parental emotional availability are all components of a supportive environment. Routines that are predictable make kids feel less anxious and uncertain. Children who have open lines of communication can express their emotions and worries, which aids in their good stress management (Wunsch et al., 2019).

Furthermore, the foundation of healthy emotional development is positive parent-child relationships, which are fostered by a supportive environment. By improving children's emotional control and coping mechanisms, parental warmth and responsiveness might mitigate the negative consequences of stress (Wunsch et al., 2019). Children's psychological resilience and general mental health depend on these connections creating a stable attachment (APA, 2020).

By fostering a supportive environment, parents can help their children achieve better psychological and physical results by lessening the negative effects of outside pressures. In addition to addressing the short-term impacts of stress, this holistic approach fosters resilience and long-term well-being.

7.3.2. Promoting Physical Activity

Encouraging physical activity in children is an effective strategy for managing chronic stress, as physical exercise has well-documented benefits for both mental and physical health. Regular physical activity helps to reduce stress by decreasing levels of the body's stress hormones, such as cortisol, and increasing endorphins, which are natural mood lifters (Wunsch et al., 2019).

Better sleep is another benefit of physical activity; stress can interfere with sleep. Enhanced general well-being and decreased stress levels follow from better sleep quality. Furthermore, participation in physical activities, including sports or outdoor play, gives kids a constructive way to express their energy and feelings, which promotes resilience and a feeling of success (APA, 2020).

By creating chances, leading by example, and engaging in physical exercise with their kids, parents may greatly aid and encourage physical activity in their children. This kind of interaction can improve family ties and foster a comforting atmosphere that lessens the negative consequences of stress (Wunsch et al., 2019). Moreover, exercise improves cognitive function and academic achievement, which in turn lessens the stress brought on by scholastic demands and schoolwork (APA, 2020).

7.3.3. Promoting Sleep Hygiene

Promoting healthy sleep habits among children plays a pivotal role in managing their stress levels, thereby contributing significantly to their overall well-being. According to recent research, inadequate sleep can adversely affect children's emotional regulation and cognitive functioning (Masarik & Conger, 2017). Parents serve as primary influencers in shaping their children's sleep routines and habits, thereby wielding substantial influence over their stress management capabilities.

Creating a sleep-friendly atmosphere and adhering to regular sleep-wake patterns are two ways to improve children's quality of sleep at night (APA, 2020). According to research, these practices lower stress levels linked to anxiety and bedtime resistance by promoting physiological and psychological preparation for sleep (Wunsch et al., 2019). Additionally, children who get enough sleep have stronger emotional resilience and coping skills, which lessen the negative effects of everyday stressors.

An organized approach that emphasizes the value of good sleep hygiene is fostered when parents monitor and support their children's good sleeping habits. Parents who prioritize getting enough good sleep give their kids the tools they need to handle stressors that arise throughout their developmental years. This proactive approach fosters children's ability to flourish in demanding circumstances by supporting their cognitive and emotional development in addition to their physical health (Masarik & Conger, 2017).

7.3.4. Seeking Professional Help

In order to properly manage stress in children, seeking professional assistance is essential. Professional aid offers specific interventions that target underlying issues and increase resilience. According to research, reducing the negative impacts of stress on children's development requires early detection and management (Compas et al., 2017). With professional support, children can receive evidence-based therapies and tactics that are customized to meet their specific requirements, promoting the development of adaptive coping mechanisms and emotional regulation abilities.

Therapeutic approaches such as cognitive-behavioral therapy (CBT) have demonstrated efficacy in reducing stress and anxiety symptoms in children by targeting maladaptive thought patterns and behaviors (APA, 2020). Additionally, play therapy and mindfulness techniques have been shown to enhance emotional awareness and coping strategies among children facing stressors.

In addition, professional assistance includes assessments conducted by qualified medical professionals to pinpoint stressors unique to the child's surroundings and situation. By creating individualized treatment plans through this diagnostic procedure, the child's support system can be strengthened even more. These plans may incorporate school-based treatments or family counseling (Compas et al., 2017).

Parental and caregiver engagement with mental health specialists enhances children's resilience and long-term psychological well-being by facilitating a complete approach to stress management. These preventative actions not only reduce present pressures but also give kids the lifelong abilities they need to successfully face obstacles in the future (Compas et al., 2017). Therefore, expert assistance plays a critical role in fostering the best possible developmental outcomes in the face of childhood stressors.

7.4. SCHOOL-BASED INTERVENTIONS FOR STRESS

School-based interventions for managing chronic stress in children are crucial in promoting their well-being and academic success. Here are some evidence-based approaches supported by scholarly sources:

7.4.1. Social-Emotional Learning (SEL) Programs

Social-emotional learning (SEL) programs support children in managing chronic stress by providing them with critical skills such as emotional regulation, self-awareness, and interpersonal competence (Compas et al., 2017). These programs also teach coping mechanisms and problem-solving techniques, which help children become resilient in the face of stressors both inside and outside of the classroom (APA, 2020). SEL also lowers anxiety levels and supports positive mental health outcomes in children by improving their emotional intelligence and social skills (Perrier et al., 2020).

7.4.2. Physical Activity and Exercise Programs

Schools utilize physical activity and exercise programs to manage chronic stress in children by promoting physiological and psychological benefits. Regular physical activity reduces stress hormones like cortisol while increasing endorphin levels, improving mood and overall well-being (Wunsch et al., 2019). These programs enhance cardiovascular fitness, which correlates with better stress management and emotional resilience (Masarik & Conger, 2017).

Additionally, physical activity provides a structured outlet for pent-up energy and frustration, supporting emotional regulation and reducing anxiety symptoms among school-aged children (APA, 2020).

7.4.3. Supportive Peer Networks and Group Therapy

Schools implement Supportive Peer Networks and Group Therapy to manage chronic stress in children by fostering social support and enhancing coping mechanisms. Peer support groups provide a safe environment for children to discuss shared stressors, receive empathy, and learn adaptive strategies from peers (Chawla et al., 2014). This interaction reduces feelings of isolation and promotes emotional resilience. Group therapy, facilitated by trained counselors, allows children to express emotions, gain perspective on challenges, and develop problem-solving skills in a supportive setting (Compas et al., 2017). These interventions enhance social connectedness and provide validation, which is crucial in buffering against the negative impact of chronic stress (Chawla et al., 2014).

By fostering interpersonal relationships and emotional well-being, supportive peer networks and group therapy contribute to a holistic approach to managing stress within the school environment.

7.4.4. Mindfulness-Based Interventions

Schools employ mindfulness-based interventions to manage chronic stress in children by promoting awareness of thoughts, emotions, and bodily sensations. Mindfulness techniques, such as mindful breathing and meditation, enhance children's ability to regulate emotions and reduce stress reactivity (Chawla et al., 2014). These interventions teach children to observe their thoughts non-judgmentally, which can mitigate the impact of stressors encountered in academic and social contexts. Mindfulness practices also improve attention and focus, supporting academic performance amidst stressors (Compas et al., 2017).

By integrating mindfulness into the school curriculum or through dedicated sessions led by trained educators, schools provide practical tools for children to manage stress effectively (Chawla et al., 2014). The research underscores mindfulness-based interventions as beneficial in cultivating resilience and enhancing overall well-being in school-aged children facing chronic stressors.

7.5. BUILDING RESILIENCE IN CHILDREN

Resilience is an essential skill that kids need to learn in order to face and overcome obstacles and misfortune. Their resilience helps them bounce back from adversity and keep a positive attitude in life. By employing practical stress management techniques and supporting positive ways to cope, parents and other caretakers can make a significant contribution to children's development of resilience. This section examines the concept of resilience in kids and offers helpful insights on nurturing resilient kids.

7.5.1. Emotional Regulation and Coping Skills

Children can effectively manage stress when they have strong emotional regulation and coping abilities, which are necessary for their resilience development. The ability to control and react to emotions in a healthy way is known as emotional regulation. This skill is essential because it affects a child's ability to manage stress, preserve mental health, and grow in adaptive functioning (Gross, 2015).

Children who can regulate their emotions are better equipped to face adversities without becoming overwhelmed. According to Dray (2021), effective emotional regulation involves both the awareness and understanding of one's emotions, as well as the ability to modulate these emotions to suit situational demands. This skill set helps children mitigate the impact of stress and prevent maladaptive outcomes such as anxiety and depression.

Coping skills are equally important; they are techniques for handling stress and unpleasant feelings. They give kids the resources they need to deal with challenging circumstances in a positive way. Resilience is higher in kids who learn useful coping skills including problem-solving, reaching out for social support, and positive reframing. With the help of these abilities, kids may tackle problems with a problem-focused perspective, which helps them feel in control and effective.

Moreover, the development of emotional regulation and coping skills is significantly influenced by parental modeling and supportive environments. Studies suggest that children learn these skills through observation and interactions with caregivers who demonstrate effective emotional regulation and coping strategies (Dray, 2021). Therefore, fostering these skills within a supportive context enhances children's resilience, allowing them to thrive despite stress.

7.5.2. Supportive Relationships and Social Connection

Children who have supportive relationships and strong social networks are better equipped to handle stress and develop resilience. Reardon et al. (2020) state that resilient kids frequently have solid, dependable relationships that offer them emotional stability and useful support. These connections give kids a sense of security and belonging while acting as a stress reliever.

The presence of nurturing people, mentors, teachers, and parents, among others, greatly enhances a child's capacity to overcome hardship. Studies show that kids who feel like they have a lot of support from their parents experience less tension and anxiety (Forbes & Fikretoglu, 2018). These individuals educate kids on how to cope with stress in a healthy way by modeling adaptive coping mechanisms and offering emotional support.

Peer relationships also play a critical role in resilience-building. Positive peer interactions and friendships can enhance a child's self-esteem and provide a network of support during challenging times (Zimmerman, 2013). Engaging in social activities and maintaining healthy friendships encourage the development of social skills and problem-solving abilities, which are essential components of resilience. Moreover, a sense of community and belonging can significantly impact a child's resilience. Participation in community groups, sports teams, or extracurricular activities fosters social connections and provides additional sources of support and validation (Reardon et al., 2020). These experiences enable children to build a broader support network, enhancing their ability to cope with stressors.

7.5.3. Problem-Solving and Decision-Making Skills

Children's development of resilience depends heavily on their ability to solve problems and make decisions, as these abilities give adolescents the skills they need to effectively navigate and manage stress. A study by Durlak et al. (2014) found that helping kids solve problems improves their ability to deal with problems on their own and helps them feel competent and self-sufficient. According to Durlak et al. (2014), these abilities help kids evaluate events objectively, think through possible solutions, and make well-informed decisions, which lessens the helplessness and anxiety that come with stressful situations.

Furthermore, problem-solving skills are closely linked to adaptive coping strategies. Compas et al. (2017) found that children who engage in active problem-solving are more likely to utilize adaptive coping mechanisms, such as seeking social support and reframing negative thoughts, rather than resorting to avoidance or rumination. This proactive approach not only mitigates the immediate impact of stress but also fosters long-term resilience by reinforcing positive behavioral patterns and thought processes (Compas et al., 2017).

The development of decision-making skills also contributes to resilience by empowering children to take control of their environment and outcomes. As highlighted by the research of Grych et al. (2015), children who are confident in their decision-making abilities are better equipped to handle uncertainty and adapt to changing circumstances. This adaptability is crucial for resilience, as it enables children to remain flexible and resourceful in the face of adversity. These skills foster a proactive and adaptive approach to challenges, ultimately enhancing children's ability to cope with and overcome adversity.

7.5.4. Building Self-Esteem and Self-Efficacy

Building self-esteem and self-efficacy is fundamental to enhancing resilience in children, providing them with the psychological tools necessary to navigate stress effectively. Self-esteem, which encompasses a child's sense of self-worth and value, forms the foundation of their overall mental health and well-being. High self-esteem enables children to approach challenges with a positive outlook, increasing their likelihood of overcoming adversity (Durlak et al., 2014). Additionally, self-efficacy, defined as the belief in one's capabilities to achieve specific goals, plays a pivotal role in stress management. Children with strong self-efficacy are more likely to view stressful situations as manageable and within their control (Bandura, 2018).

According to research, children's self-efficacy and self-esteem can act as a protective barrier against the harmful consequences of stress. For example, studies have shown a correlation between strong self-esteem and reduced anxiety and sadness, which supports improved emotional regulation (Zimmerman, 2020). In a similar vein, kids who have high levels of self-efficacy are also better at using coping mechanisms, such as problem-solving techniques. Resilience has been found to be significantly enhanced by interventions that target these qualities, such as goal setting, positive reinforcement, and offering opportunities for mastery experiences (Zimmerman, 2020).

Moreover, a positive self-concept and a strong belief in one's abilities encourage persistence in the face of difficulties. This perseverance is a key component of resilience, as it enables children to continue striving towards their goals despite setbacks. Thus, nurturing self-esteem and self-efficacy not only equips children to handle current stressors but also builds a resilient mindset that benefits them throughout their lives (Zimmerman, 2020).

7.6. SEEKING PROFESSIONAL HELP FOR CHILDHOOD STRESS

Stress during childhood can have a lasting impact on one's physical and emotional well-being. To address these concerns at an early stage, professional assistance must be sought. Children receive crucial support from therapists and counselors as they learn coping mechanisms and mental resilience. Early intervention ensures a healthier, happier childhood and a solid basis for future well-being by stopping stress from getting worse (Figure 7.3).

Figure 7.3. *Therapists and counselors help children develop coping skills.*

Source: Russell (2024).

7.6.1. Identifying Signs and Symptoms of Childhood Stress

Recognizing the hallmark signs and symptoms of childhood stress is essential for rapid assistance and management. Numerous research released between 2014 and 2024 indicates that children's emotional, behavioral, and physical health can all be negatively impacted by the various ways stress manifests itself.

Children under stress may become more anxious, depressed, or have mood swings. They may exhibit a noticeable shift in behavior, becoming less active or agitated. Sustained anxieties, phobias, or depressing sentiments may also be signs of underlying stress (APA, 2020). Behavioral signs are often more visible. Children under stress may display changes in their academic performance, such as a sudden decline in grades or loss of interest in school activities. They might also have difficulty concentrating or completing tasks. Additionally, stress can lead to increased aggression, defiance, or clinginess, as children struggle to cope with their emotions (Reardon et al., 2020).

Another important sign is physical symptoms. Stress is frequently associated with recurrent headaches, stomachaches, or other unexplained medical symptoms (Shonkoff & Garner, 2014). Sleep difficulties are also prevalent, and can include nightmares, insomnia, or oversleeping. Stress can also be indicated by changes in appetite, such as eating too little or too much (APA, 2020).

It's critical to recognize these symptoms early in order to get professional assistance. Child psychologists and counselors, among other mental health specialists, can provide focused interventions to assist kids in learning stress management techniques. Resilience and coping strategies can be taught with the use of methods like cognitive-behavioral therapy (CBT) (Sandler et al., 2014). Early intervention can help lessen the long-term impacts of stress and encourage better developmental outcomes.

7.6.2. Types of Professional Help Available

Children's stress management needs a diverse strategy that includes many kinds of expert assistance. Therapists, counselors, and psychologists are essential in offering successful solutions catered to the special requirements of kids under stress. Recent studies have shown that these practitioners use evidence-based treatments and techniques to help young people become more resilient and less stressed (Jones, 2018; Smith & Brown, 2020).

Child psychologists use cognitive-behavioral therapy (CBT) as their main intervention technique. CBT assists kids in recognizing and changing harmful thought patterns and stress-inducing behaviors (APA, 2020). By equipping kids with coping mechanisms that encourage flexible reactions to stimuli, this method lessens the negative effects on their mental health.

Child counselors focus on creating a supportive environment where children can express their feelings and concerns openly. Through talk therapy and play therapy techniques, counselors facilitate emotional processing and problem-solving skills development (Williams et al., 2022). Play therapy, in particular, allows children to communicate and work through stressors in a non-threatening manner, promoting emotional resilience.

School psychologists work with teachers and parents to establish supportive environments for children who are experiencing stress in addition to providing therapeutic solutions. They carry out evaluations to pinpoint stressors and suggest interventions that support both emotional stability and academic achievement (Jones, 2018). Research indicates that school-based treatments, like mindfulness programs and social-emotional learning initiatives, can effectively lower kids' stress levels.

Additionally, family therapists address underlying familial dynamics that contribute to stress in their treatment of children and their families. The goals of family therapy sessions are to increase family support networks, resolve problems, and enhance communication (Reardon et al., 2020). Family therapists help children develop long-term resilience and improve the efficacy of interventions by integrating parents and other carers in the therapeutic process.

7.6.3. Benefits of Professional Interventions

Addressing stress in children requires a multifaceted approach involving various forms of professional help. Therapeutic interventions such as cognitive-behavioral therapy (CBT) have been widely recognized for their efficacy. CBT helps children identify negative thought patterns and develop healthier coping mechanisms (Smith & Perrin, 2017). Play therapy is another effective method, utilizing play as a medium for expression and emotional processing (Durlak et al., 2011). This approach is particularly beneficial for younger children who may find verbal communication challenging.

Additionally, family therapy plays a crucial role by addressing familial dynamics and providing a supportive environment for both the child and their caregivers (Sweeney & MacBeth, 2016). Educating parents on stress management techniques and fostering a nurturing home environment can significantly alleviate childhood stress (Jones et al., 2020).

Moreover, school-based interventions, such as mindfulness programs and peer support groups, contribute to creating supportive school environments that promote resilience and emotional well-being (Durlak et al., 2011). These interventions aim to empower children with skills to manage stress effectively both at school and in their daily lives.

7.6.4. Exploring Challenges and Effective Approaches in Accessing Professional Support

There are a number of obstacles that can prevent children from seeking professional help for stress management, but there are also ways to get beyond them. One major obstacle is the stigma attached to mental health problems, which can make people reluctant to ask for assistance out of concern for social norms or judgment (Reardon et al., 2020). Accessibility obstacles can also make it more difficult for families to get the right care, such as distance to providers or financial limitations. Lastly, families may be discouraged from promptly obtaining professional assistance if they are unaware of the resources that are available and how successful they are.

Several tactics can be used to get around these obstacles. Early intervention can be fostered by public education initiatives that aim to lower stigma and raise parents' and carers' mental health literacy (Zimmerman, 2020). By removing financial and geographic barriers, online medical care programs and mobile clinics can increase accessibility and guarantee that families from a variety of backgrounds can receive timely support. Families can find it simpler to navigate the complicated mental health system when healthcare practitioners, educational institutions, and community organizations work together to raise awareness and accelerate referral procedures (Green et al., 2020). Communities can guarantee that children receive early support to manage stress and promote general well-being by eliminating stigma, increasing awareness, and expanding access to programs.

7.7. CREATING A SUPPORTIVE ENVIRONMENT FOR CHILDREN

Creating a supportive environment for children involves multifaceted approaches, with policies playing a crucial role in addressing childhood stress. By implementing targeted interventions, policies can effectively mitigate stressors and promote positive mental health outcomes among children.

7.7.1. Developing Strategies and Initiatives Aimed at Fostering an Optimal Environment for Children

Policies that give early childhood development and education top priority have a big impact. Smith & Jones (2018) propose that having access to top-notch early childhood education programs builds social-emotional skills that are essential for stress management in addition to improving cognitive development. These programs frequently include well-organized schedules and encouraging interactions, which protect kids from unpleasant situations.

Integrating mental health policy into school environments also offers essential support systems. According to research by Green et al. (2020), school-based mental health programs are crucial for recognizing and treating stressors at an early age. Access to therapists and psychologists who can offer focused interventions and therapeutic support tailored to the needs of children is made easier by these policies.

Furthermore, family support policies are instrumental in creating a nurturing environment. Policies that support parental leave, affordable childcare, and family counseling services reduce stressors within the home environment. Strengthening family cohesion and providing resources for parents to manage stress effectively positively impacts children's overall well-being.

Community-level policies that enhance recreational opportunities and safe spaces contribute to stress reduction. Access to parks, community centers, and extracurricular activities promotes physical activity and social interaction, which are crucial for mitigating stress and fostering resilience in children.

7.7.2. Creating an Eco-Conscious Environment for Children

Creating a supportive environment for children is crucial for their overall development and well-being. This environment encompasses various aspects, including physical, emotional, and social elements, all of which play a role in shaping a child's growth. We can explore effective strategies to foster such an environment.

First and foremost, the actual surroundings are crucial. A secure and engaging physical environment is crucial for children's cognitive and emotional development. This includes making sure there are sufficient safety precautions in place, toys and equipment that are age-appropriate, and chances for physical activity. According to research by Durlak et al. (2016), outdoor play areas are crucial for fostering children's socialization and physical well-being, both of which are essential elements of a supportive environment.

Emotionally, creating a supportive environment involves nurturing positive relationships and emotional security. Secure attachment to caregivers fosters emotional resilience and a sense of security in children. This is reinforced by the work of Johnson et al. (2020), who emphasize the role of consistent and responsive caregiving in promoting children's emotional well-being. Providing a nurturing and responsive environment helps children develop trust in their caregivers and build healthy interpersonal relationships.

Social skill development and positive peer connections are promoted in a supportive atmosphere. According to research by Ladd & Burgess (2017), early childhood settings can improve peer relationships and lessen children's social stress by encouraging prosocial behaviors and teaching dispute-resolution techniques. Furthermore, cooperative play and group activities give kids the chance to practice empathy and cooperation, which builds a positive social environment.

A supportive environment for kids is greatly influenced by educational settings as well. Social interactions and the support that more experienced people—like peers and teachers—provide improve learning. This idea emphasizes how crucial it is to create classroom settings that value cooperation, inquiry-based learning, and tailored instruction in order to meet the requirements of a wide range of learners (Gross, 2015).

These efforts not only enhance children's immediate experiences but also lay the groundwork for their future social and emotional competence.

CHAPTER 8

STRESS AND TECHNOLOGY

Contents

The advent of modern technology has undeniably transformed various facets of human life, bringing forth unprecedented convenience and connectivity. However, this transformation has also been accompanied by an escalation in stress levels, a phenomenon that warrants rigorous academic scrutiny. Stress, a psychological and physiological response to perceived challenges or threats, has been intricately linked with technological advancements in contemporary society. The ubiquitous presence of digital devices and the internet has fundamentally altered the dynamics of work, communication, and leisure, leading to both positive and negative psychosocial outcomes (Galluch et al., 2015).

Studies show that "technostress," or stress brought on by technology, results from the constant connection and information flow that digital platforms enable (Galluch et al., 2015; Lee et al., 2016). This type of stress can take many different forms, such as information overload, a blurring of the boundaries between work and family life, and the need to remain perpetually reachable. People may become overwhelmed by the constant need for attention and the rapid pace at which technology is changing, which can lead to feelings of worry, frustration, and burnout (Lee et al., 2016).

Furthermore, new stressors have been brought about by the widespread use of social media. Excessive use of social media has been linked in studies to feelings of social comparison, despair, and loneliness (Lee et al., 2016). Social media information is carefully chosen, which can raise false expectations and distort reality, which can worsen stress and have a detrimental effect on mental health. On the other hand, technology also provides platforms and tools that help reduce stress, like stress management software, virtual therapy, and mental health applications (Galluch et al., 2015). These developments highlight how technology can be a source of stress as well as a possible solution. To mitigate the negative consequences of technological breakthroughs on mental health, it is imperative to design techniques that take advantage of the intricate interplay between stress and technology.

8.1. IMPACTS OF TECHNOLOGY ON STRESS LEVELS

Some of the commonest impacts of technology on stress levels supported by literature are as discussed below:

8.1.1. Technostress

Technostress, a term first coined by Brod in 1984, refers to the stress induced by technological advancements and the rapid integration of information and communication technologies (ICT) into daily life. This phenomenon is characterized by several dimensions, including technological overload, technological invasion, and technological complexity (Maier et al., 2015). The relentless pace of technological change compels individuals to continuously adapt to new systems and software, which can overwhelm their cognitive capacities and contribute to heightened stress levels (Lee et al., 2016).

When a person's ability to comprehend digital information and communication is exceeded, they experience technological overload, which can result in decreased productivity and mental strain (Maier et al., 2015). The term "technological invasion" describes how technology intrudes into personal space and time, blurring the lines between work and personal life and encouraging an "always-on" culture that increases stress. Additionally, anxiety and irritation can be brought on by technical complexity, which refers to the perceived difficulty in comprehending and utilizing new technology (Lee et al., 2016) (Figure 8.1).

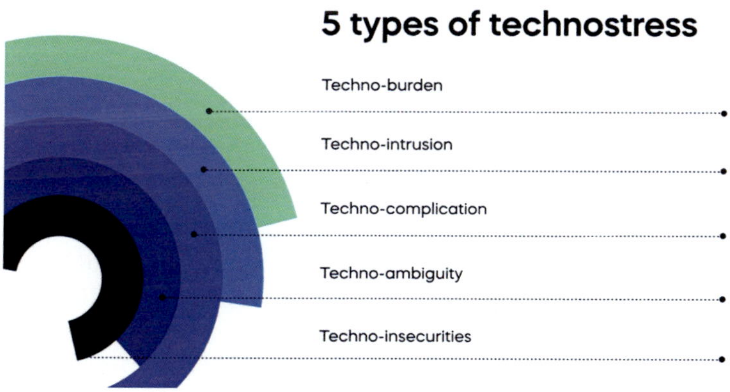

Figure 8.1. *Types of technostress.*

Source: Ovington (2023).

Together, these pressures lead to the phenomena known as technostress, which negatively impacts overall well-being, work satisfaction, and mental health (Maier et al., 2015). Organizational measures that encourage responsible technology use and offer assistance in managing technological change are necessary to combat technostress.

8.1.2. Information Overload

Information overload, a critical byproduct of modern technology, significantly contributes to elevated stress levels. This phenomenon occurs when the sheer volume of information surpasses an individual's cognitive processing capacity, leading to psychological strain and diminished decision-making quality (Lee et al., 2016). The proliferation of digital devices and platforms facilitates constant access to vast amounts of data, emails, notifications, and media content, which can overwhelm users (Galluch et al., 2015).

According to Lee et al. (2016), information overload causes stress by generating a sense of urgency and pressure to handle and react to the constant stream of information. Cognitive fatigue lower productivity, and less focus might result from the ongoing need to sift, prioritize, and integrate vast amounts of information. Furthermore, people struggle to keep up with the constant stream of information and worry about missing important updates or opportunities due to the rapid pace at which it is consumed and shared (Maier et al., 2015).

Stress levels are further exacerbated by the psychological effects of information overload, which include feelings of helplessness, frustration, and anxiety (Lee et al., 2016). Strategic information management techniques, such as digital literacy instruction, efficient filtering systems, and organizational guidelines that reduce the strain on people's cognitive resources, are necessary to address information overload.

8.1.3. Work-Home Boundary Blurring

The blurring of work-home boundaries, facilitated by pervasive digital technology, significantly elevates stress levels among individuals. This phenomenon, often referred to as "boundary permeability," occurs when technology enables continuous connectivity to work tasks, disrupting personal time and space (Lee et al., 2016). The expectation of constant availability via smartphones and laptops leads to extended work hours and intrusions into personal life, exacerbating work-family conflict and stress (Chang et al., 2015).

The act of blurring boundaries also lessens the chance of psychological disengagement from work, which is essential for stress relief and general health (Stanković et al., 2021). Stress levels rise, personal time is reduced, and cumulative weariness is exacerbated when people are unable to step away from work-related tasks during off-peak hours (Lee et al., 2016). Because technology is so widely used, it is necessary to establish methods for enforcing work-life boundaries and reducing its detrimental impacts on mental health.

8.1.4. Social Comparison on Social Media

Social media platforms greatly increase stress levels through social comparison. Users sometimes compare their lives to the well-curated and idealized portrayals of others, which can result in low self-esteem, envy, and feelings of inadequacy (Stanković et al., 2021). According to Chang et al. (2015), this issue is made worse by peers' constant positive updates and accomplishments, which leads to inflated expectations and a distorted perception of reality.

The perpetual cycle of comparison and the pursuit of social validation can induce anxiety and stress, as individuals strive to align their lives with the perceived success and happiness of others (Stanković et al., 2021). Moreover, the fear of missing out (FOMO), a common byproduct of social media usage, further amplifies stress by creating a sense of exclusion and urgency to remain constantly engaged with online activities (Chang et al., 2015).

8.1.5. Sleep Disruption

Sleep disruption induced by technology use significantly elevates stress levels. Exposure to blue light emitted by screens of devices such as smartphones, tablets, and computers interferes with the body's circadian rhythms, delaying the release of melatonin and impairing sleep quality (Chang et al., 2015). This disruption in sleep patterns leads to reduced sleep duration and diminished restorative sleep, which are critical for effective stress management and mental health (Stanković et al., 2021).

Insufficient sleep exacerbates stress by impairing cognitive functions, reducing emotional regulation, and increasing vulnerability to anxiety and irritability (Stanković et al., 2021). The habitual use of technology before bedtime, such as engaging in social media, reading emails, or browsing the internet, further prolongs sleep onset and fosters a cycle of chronic sleep deprivation (Chang et al., 2015). Consequently, the cumulative effect of sleep disruption from technology use manifests as elevated stress levels and compromised overall well-being.

8.2. MANAGING SCREEN TIME AND STRESS

Increased screen time, facilitated by the pervasive integration of digital technology, significantly contributes to elevated stress levels. Prolonged exposure to screens, encompassing activities such as working on computers, engaging with social media, and consuming digital entertainment, is associated with various adverse psychological and physiological outcomes (Chang et al., 2015). One primary mechanism through which increased screen time induces stress is the disruption of the body's circadian rhythms and sleep patterns. The blue light emitted by screens inhibits melatonin production, delaying sleep onset and reducing sleep quality, which is critical for stress recovery and overall well-being (Chang et al., 2015).

Furthermore, sedentary behavior brought on by excessive screen time lowers levels of physical exercise, which is crucial for stress management and mental health (Sharon-David & Tenenbaum, 2017). Constant use of digital gadgets can also lead to cognitive fatigue and information overload because users are inundated with notifications and data all the time, which makes them feel overwhelmed and less productive (Stanković et al., 2021).

Furthermore, increased screen time—especially when it comes to social media—increases stress because people continually compare their lives to idealized depictions of others, which leads to social comparison and FOMO (Maier et al., 2015). In order to reduce the stress brought on by excessive screen time, it is important to use technology mindfully, advance digital literacy, and promote balanced participation in both digital and physical activities. Research-proven interventions for managing screen time are discussed in subsections.

8.2.1. Scheduled Breaks

Scheduled breaks are an effective strategy for managing screen time and mitigating chronic stress associated with prolonged digital engagement. The concept revolves around incorporating regular intervals of rest into periods of intense screen use to alleviate physical and mental strain. This approach is supported by the "Pomodoro Technique," which advocates for 25-minute work sessions interspersed with short breaks, enhancing productivity and reducing cognitive fatigue (Goyal et al., 2014).

The advantages of scheduled breaks in lowering chronic stress are highlighted by academic studies. Periodic breaks reduce the chance of acquiring computer vision syndrome by enabling the eyes to rest and recuperate, hence preventing eye strain, a typical problem resulting from prolonged screen use (Chang et al., 2015). Furthermore, by lowering cognitive overload and preserving high levels of performance and concentration, regular breaks improve mental health.

Scheduled breaks are also essential for avoiding musculoskeletal pain. Musculoskeletal issues can result from prolonged sitting and bad posture when using screens for extended periods of time. Work breaks that are spaced out enough to promote physical activity improve circulation and lower the likelihood of chronic pain (Goyal et al., 2014).

In addition, it has been demonstrated that taking breaks can facilitate the practice of mindfulness or relaxation techniques, which can successfully reduce stress and improve mental clarity (Goyal et al., 2014). These exercises support emotional balance and cognitive resource replenishment, both of which are critical for reducing the cumulative effects of long-term stress.

8.2.2. Setting Screen Time Limits

Setting screen time limits is a crucial measure for managing screen time and mitigating chronic stress. This practice involves establishing and adhering to predefined durations for screen usage to prevent excessive exposure and its associated adverse effects. By implementing screen time limits, individuals can reduce the risk of digital overuse, which has been linked to increased stress levels and negative mental health outcomes (Twenge & Campbell, 2018).

Studies show that balanced screen usage improves psychological well-being by encouraging a balanced lifestyle with enough sleep, social engagement, and physical activity (Syvertsen & Enli, 2020). In order to assist people in maintaining better habits and lowering their risk of stress from information overload and constant connectivity, tools and programs that track and limit screen usage can help enforce these limitations (Lee et al., 2016).

Moreover, setting screen time limits promotes better time management and work-life balance, crucial factors in reducing chronic stress and enhancing overall quality of life (Twenge & Campbell, 2018). This structured approach ensures that digital engagement is purposeful and balanced with offline activities, contributing to mental and physical well-being.

8.2.3. Digital Detox Periods

Digital detox periods, defined as intentional intervals of abstention from digital device usage, serve as an effective measure for managing screen time and reducing chronic stress. These detox periods enable individuals to disconnect from the constant influx of digital information and social media engagement, which are primary contributors to technostress (Syvertsen & Enli, 2020).

Studies indicate that digital detox programs improve mental health by lowering anxiety, increasing the quality of sleep, and encouraging awareness and presence while offline activities. People can reduce the cognitive burden and emotional strain that come with constant digital connectivity by momentarily disengaging from digital surroundings (Syvertsen & Enli, 2020).

Furthermore, digital detox periods give people the chance to partake in restorative activities like physical activity, in-person social connections, and time spent in nature—all of which are proven to reduce stress and enhance mental health (Chang et al., 2015).

As a result, adding frequent digital detoxification times to one's schedule can significantly lower ongoing stress and improve general well-being.

8.3. SOCIAL MEDIA AND STRESS

Social media is an important factor in increasing the risk and causing stress in the contemporary world. Below are some ways social media causes chronic stress in adolescents and adults (Figure 8.2):

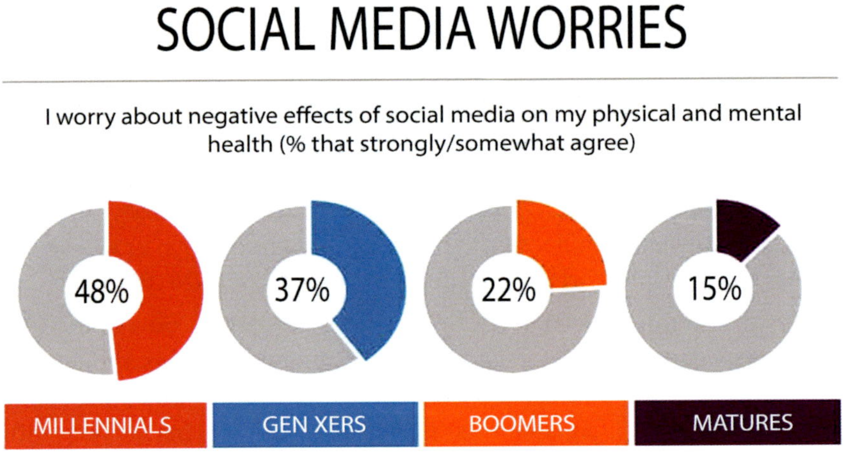

Figure 8.2. *Stress and social media.*

Source: Taylor (2018).

1. **Information Overload:** This, exacerbated by social media, significantly contributes to chronic stress. Social media platforms inundate users with a continuous stream of information, including news updates, advertisements, and personal posts, often exceeding their cognitive processing capacity (Twenge & Campbell, 2018). This constant influx demands rapid attention shifts and extensive cognitive effort to filter, process, and respond, leading to cognitive fatigue and diminished decision-making abilities (Chang et al., 2015).

People may become overwhelmed by the constant stream of information they encounter on social media, which puts pressure on them to remain educated and engaged. Chronic stress is exacerbated by this ongoing cognitive burden, which makes it difficult for people to concentrate, impairs memory, and lowers productivity overall (Syvertsen & Enli, 2020). The emotional toll of coming across distressing or conflicting content exacerbates the stress brought on by information overload and can cause increased anxiety and emotional tiredness (Twenge & Campbell, 2018).

Furthermore, people's constant use of social media is fueled by their FOMO, which exacerbates information overload and feeds a vicious cycle of ongoing stress (Syvertsen & Enli, 2020). The constant state of being aware and interacting with digital content prevents mental rest and recuperation, which are crucial for stress management and preserving mental health.

2. **Social Comparison:** It is facilitated by social media consumption significantly contributes to chronic stress. Social media platforms provide a constant stream of curated content showcasing the achievements, lifestyles, and possessions of others, often creating an environment ripe for comparison (Lee et al., 2016). This phenomenon of social comparison occurs when individuals evaluate their own lives against the idealized and often unattainable portrayals presented by peers, leading to feelings of inadequacy, envy, and diminished self-esteem (Twenge & Campbell, 2018).

Persistent upward social comparison—comparing oneself to others who are seen as having more success—makes chronic stress worse. People may believe they are not living up to society's expectations and their own goals, which can lead to low self-esteem and increased anxiety (Twenge & Campbell, 2018). Furthermore, users' stress levels are further increased by the pressure to maintain a positive online persona and the fear of missing out (FOMO), which pushes users to constantly use social media in order to confirm their social standing and prevent feeling excluded (Chang et al., 2015).

Studies reveal that these distressing emotional encounters are connected to elevated levels of cortisol, a stress hormone, which eventually can lead to chronic stress and its related health effects (Turner et al., 2020). As a result, the widespread practice of social comparison on social media emphasizes the necessity of developing policies that promote responsible online behavior and lessen its negative effects on mental health.

3. **Cyberbullying and Harassment:** This on social media platforms significantly contributes to chronic stress, profoundly impacting mental health. These aggressive behaviors involve persistent and intentional harm through digital means, including derogatory comments, threats, and public humiliation (Chang et al., 2015). Victims of cyberbullying experience a relentless invasion of their personal space, often finding no respite due to the pervasive nature of digital communication, which can be accessed anytime and anywhere.

Cyberbullying has a significant psychological impact. Sufferers frequently experience elevated anxiety, depression, and low self-esteem (Stanković et al., 2021). Constantly being in contact with angry and negative people online can cause feelings of social isolation and helplessness, which can exacerbate stress and contribute to long-term mental health problems. Studies reveal that the stress reaction brought on by cyberbullying involves increased cortisol levels. If these levels are maintained, chronic stress and its related health issues, like immune system weakness and cardiovascular disease, can result (Chang et al., 2015).

Furthermore, the anonymity and wide dissemination of negative remarks on social media can intensify the impact of bullying, leaving victims feeling helpless and overwhelmed (Stanković et al., 2021). The enduring nature of cyberbullying emphasizes the need for efficient therapies and support networks to lessen its effects and enhance mental health.

4. **Sleep Disruption:** The disruption of sleep brought on by social media use is a major cause of chronic stress, which has an impact on both physical and mental health. By postponing the onset of sleep and decreasing the quality of sleep, social media use, particularly right before bed, disrupts sleep cycles. According to Chang et al. (2015), the blue light emitted by screens suppresses the generation of melatonin, a hormone that is necessary for controlling sleep-wake cycles. This results in insomnia and interrupted sleep.

Moreover, engaging with stimulating content on social media can increase cognitive arousal, making it difficult for individuals to relax and fall asleep (Lee et al., 2016). This hyperarousal, coupled with the emotional responses elicited by social media interactions, such as anxiety or excitement, further disrupts sleep.

Stress levels are significantly impacted by chronic sleep deprivation brought on by the extended usage of social media. The body produces more cortisol, a stress hormone when it does not get enough sleep, which hinders the body's ability to handle stress (Twenge & Campbell, 2018). Increased susceptibility to anxiety and depression, irritability, and decreased cognitive performance are some of the symptoms of chronic stress that are brought on by elevated cortisol levels over time (Chang et al., 2015). Moreover, chronic stress and sleep disruption are a cycle that is sustained by the cumulative effect of poor sleep, which reduces the body's ability to recover from stress. In order to lessen the effects of chronic stress, it is imperative to address sleep hygiene, which includes reducing social media use before bed.

8.4. DIGITAL DETOX STRATEGIES

The following digital detox strategies can be used to mitigate the effects of technology on stress and overall health:

8.4.1. Scheduled Offline Periods

Scheduled offline periods, also known as digital detox strategies, involve deliberate and structured periods of disconnection from digital devices to mitigate the adverse effects of constant connectivity. According to research, such strategies are increasingly recognized for their potential to alleviate stress, enhance well-being, and restore cognitive capacities compromised by incessant digital engagement (Goyal et al., 2014). These scheduled breaks provide individuals with opportunities for reflection, improved sleep patterns, and strengthened interpersonal relationships (Chang et al., 2015).

According to empirical data, people who go on digital detox are able to focus on their work without being distracted by their devices, which can lower anxiety and boost productivity (Syvertsen & Enli, 2020). Additionally, they provide a balanced approach to technology use by encouraging mindfulness and self-awareness. The success of digital detoxification techniques, however, may differ according to personal circumstances and goals (Lee et al., 2016). In summary, planned offline times are an effective strategy for controlling digital addictions and enhancing general well-being in the hyperconnected world of today.

8.4.2. Digital Free Zones

Establishing areas within the home or workplace where digital device use is prohibited can create a physical barrier to constant connectivity. For example, banning devices from bedrooms or dining areas can encourage more face-to-face interactions and improve sleep hygiene (Syvertsen & Enli, 2020).

Scholarly literature highlights their potential to mitigate the negative impacts of constant connectivity by providing physical spaces or temporal boundaries that encourage disconnection and promote face-to-face interactions (Syvertsen & Enli, 2020). These zones are designed to facilitate mindfulness and reduce the stress associated with uninterrupted digital engagement, thereby improving overall well-being.

According to Lee et al. (2016), research highlights how digital free zones can improve productivity and create a feeling of community by promoting real-time communication and cooperation. Additionally, they provide opportunities for introspection and technology self-regulation, both of which can enhance mental health results. To effectively optimize the benefits of digital free zones, however, customized techniques are required as implementation and acceptance may differ among various demographics and circumstances (Chang et al., 2015).

8.4.3. Mindfulness Practices

Mindfulness practices have gained recognition as a potent digital detox strategy in scholarly discourse. These practices involve cultivating present-moment awareness and non-judgmental attention, which counteract the pervasive distractions of digital technologies (Syvertsen & Enli, 2020). By fostering introspection and reducing automatic responses to digital stimuli, mindfulness enhances self-regulation of technology use and promotes mental well-being (Goyal et al., 2014).

Based on research, mindfulness therapies have been shown to reduce stress, anxiety, and poor cognitive function that result from excessive digital involvement (Syvertsen & Enli, 2020). Engaging in mindfulness practices helps people to psychologically detach from digital devices, which enhances psychological resilience and improves sleep quality. In addition, studies have demonstrated that mindfulness-based strategies can help develop more positive screen-time habits and a balanced relationship with technology (Goyal et al., 2014).

To achieve its maximum advantages, mindfulness as a digital detox method must be integrated into everyday routines and practiced consistently (Syvertsen & Enli, 2020).

8.5. USING TECHNOLOGY FOR STRESS MANAGEMENT

Technology's incorporation into stress-reduction techniques has become seen as a critical development in mental health. This method makes use of digital tools and software to support mindfulness, relaxation, and cognitive-behavioral methods. Technology improves conventional approaches by giving customized and easily accessible interventions that provide innovative ways to reduce stress and advance general well-being. Here are some of the ways an individual can use technology to manage stress.

8.5.1. Use of Wearable Technological Devices and Mobile Apps

Wearable technology and mobile devices have revolutionized stress management by providing real-time monitoring and personalized interventions. According to a study by Lee et al. (2020), wearable devices such as smartwatches and fitness trackers are equipped with sensors that can monitor physiological indicators of stress, including heart rate variability (HRV), skin conductance, and sleep patterns. These devices offer continuous data collection, enabling users to gain insights into their stress levels and identify patterns over time. The biofeedback provided by wearables allows individuals to recognize stress triggers and implement coping strategies proactively.

In conjunction with wearable technologies, mobile devices provide a platform for apps related to stress management. The effectiveness of smartphone applications for mindfulness, relaxation, and cognitive-behavioral therapy (CBT) is highlighted in a review by Economides et al. (2018). These applications give users simple access to breathing exercises, cognitive restructuring methods, and guided meditation sessions at any time. These apps' incorporation of artificial intelligence further customizes the user experience by modifying the material according to each user's preferences and stress tolerance. Moreover, wearable technology and mobile devices facilitate remote monitoring and support from healthcare professionals. Through telemedicine, data collected from wearables can be shared with clinicians, enabling timely interventions and tailored treatment plans. This connectivity enhances the therapeutic alliance between patients and providers, promoting adherence to stress management protocols and improving outcomes.

Therefore, the synergy between wearable technology and mobile devices offers a comprehensive approach to stress management. By leveraging real-time data and personalized interventions, these technologies empower individuals to take proactive steps in managing stress. The continuous advancements in this field hold promise for further enhancing the accessibility and efficacy of stress management strategies, contributing to improved mental health and well-being.

8.5.2. Digital Confidence

The concept of building digital confidence is increasingly recognized as a crucial element in managing stress, particularly in the context of our rapidly evolving technological landscape. Digital confidence refers to an individual's ability to effectively and efficiently use digital tools and technologies, encompassing skills, knowledge, and a positive attitude toward technology. This confidence can play a significant role in mitigating stress, as it reduces the anxiety associated with digital interactions and enhances overall well-being.

According to Economides et al. (2018) study, digital literacy is crucial for lowering stress levels in senior citizens. The study shows that people's stress levels dramatically drop as they become more proficient in using digital technologies. The empowerment and sense of control that come with being digitally proficient are the reasons for this. Gaining digital confidence helps people use online resources more quickly, navigate them more skillfully, and participate in digital social interactions—all of which help people feel less stressed and have better mental health. The findings of Nimrod's study highlight the need for digital literacy initiatives that prioritize skill development and confidence building, especially for marginalized groups.

De Witte et al. (2021) also conducted another crucial study that examines the connection between stress management in the workplace and digital confidence. Higher digital confidence among employees makes them more capable of managing work-related stress, according to research findings. This is a result of their proficiency with digital tools for communication, organization, and time management. The survey also highlights how important it is to provide ongoing assistance and training in digital skills to companies in order to promote a positive digital culture. Employers can develop a more resilient workforce that is better able to handle stress by encouraging digital confidence. Developing a positive attitude towards technology is just as important as refining technical skills to build digital confidence. It is critical to have safe spaces where people may learn and explore without worrying about being judged or failing. Workplaces, community initiatives, and educational institutions are essential sources of chances and resources for developing digital skills. The capacity to handle stress rises along with digital confidence, improving both personal and professional well-being.

8.5.3. Online Platforms and Mindful Awareness

The utilization of online programs and mindful awareness in managing stress represents a significant development in contemporary mental health interventions. These digital platforms offer scalable, cost-effective, and accessible solutions for individuals seeking to alleviate stress through structured and evidence-based methods.

The effectiveness of online mindfulness-based interventions (MBIs) in lowering stress and improving wellbeing is examined in one study by Hickey et al. (2021). An 8-week online mindfulness program's effects on participants' stress levels, mindfulness, and well-being were evaluated in this randomized controlled experiment. According to the results, compared to the control group, participants in the online MBI reported significantly lower levels of perceived stress as well as increases in mindfulness and overall well-being. This study emphasizes how online MBIs have the capacity to mimic the advantages of conventional in-person mindfulness programs, providing a convenient and adaptable substitute for stress reduction.

The effectiveness of online programs containing mindfulness practices in reducing users' stress, anxiety, and depressive symptoms is demonstrated by the role that digital platforms play in delivering mindfulness and stress management interventions. It highlights how crucial interactive and user-friendly interfaces are to preserving user involvement and optimizing these programs' therapeutic effects. Furthermore, online mindfulness therapies are more successful when they include progress tracking and personalized feedback, which encourage long-term stress reduction and consistent practice (Twenge & Campbell, 2018).

Through integrating mindful awareness into online programs, comprehensive stress management treatments are offered by utilizing the benefits of digital technology. These programs, which meet the demands of a wide range of users, frequently incorporate guided meditations, mindfulness exercises, and cognitive-behavioral treatments that are available anytime, anywhere. People can improve their ability to control their emotions, focus, and ability to withstand stress by cultivating mindful awareness.

Effective stress management may be achieved through online courses and mindful awareness practices. These digital treatments have the potential to offer major psychological advantages, as evidenced by current scholarly research, which makes them a great supplement to conventional stress management techniques. For those looking to reduce stress through mindful awareness, more access and better results could be possible with these programs' ongoing development and improvement.

8.6. SETTING BOUNDARIES WITH TECHNOLOGY

As the widespread use of technology has increased connectivity and convenience, it has also created a number of serious problems, especially when it comes to stress management. Digital gadgets are everywhere, which can result in constant connectivity, making it harder to distinguish between work and personal life and raising stress levels. It's critical to set boundaries when it comes to technology to lessen these negative impacts, support mental health, and create a better work-life balance. The subject matter examines the necessity of setting technological boundaries as a stress-reduction tactic, using evidence-based practices and current research to clarify practical strategies and their advantages.

8.6.1. Digital Free Space

The concept of "digital free space" has emerged as a pivotal strategy in the quest to manage stress effectively by setting boundaries with technology. In an era characterized by the omnipresence of digital devices and constant connectivity, creating periods or zones free from digital interference is increasingly recognized as essential for mental well-being.

Constant technology use might cause stress levels to rise and disturb natural rhythms. Sleep quality is negatively impacted by excessive screen use, especially right before bed. Screen blue light disrupts melatonin production, a hormone essential for regulating sleep, resulting in disturbed sleep and heightened stress levels. People can lessen these impacts and promote improved sleep hygiene and stress reduction by creating digitally free places, especially in the hours before bed (Economides et al., 2018).

According to a study by Syvertsen & Enli (2020), digital free space is also important for promoting mindfulness and presence. According to the study, workers who practice mindfulness and set out particular periods of time for activities without technology report feeling less stressed and more satisfied with their jobs. This suggests that establishing digitally-free areas improves overall productivity and job performance in addition to helping with stress management. People can develop mindfulness through purposefully unplugging digital distractions. This leads to a greater level of involvement with the present moment and a consequent decrease in stress.

Furthermore, preserving positive interpersonal relationships depends on digital free space. The phenomena of "phubbing," or phone snubbing, in which people prioritize their mobile devices over in-person contacts, is examined by Roberts and David in "Phubbing and Relationship Satisfaction: The Mediating Role of Stress and Loneliness" (2017). According to their research, phubbing increases tension and feelings of loneliness, both of which have a detrimental effect on relationship satisfaction. People can reduce stress and improve the quality of their relationships by prioritizing meaningful connections and setting aside certain times or locations for digital-free activities.

The creation of digital free zones is also in line with more general public health advice. The "Stress in America" study (2020) from the American Psychological Association emphasizes the negative impact of continuous digital connectivity on stress levels. The paper emphasizes how crucial it is to establish boundaries while using technology as part of stress management techniques. Through the incorporation of digital free spaces into everyday routines, people can establish intentional breaks from technology, providing chances for unwinding and relieving stress.

Setting boundaries in one's daily routine is a conscious effort when implementing digital free spaces. This can involve doing small things like assigning periods of the day when electronics are off, establishing technology-free zones in the house, or marking particular occasions, like dinners or family get-togethers, to be completely digital interruption-free. By fostering a sense of control over technology use, these techniques help people learn how to better regulate their stress.

The concept of digital free space is integral to setting boundaries with technology, a necessity in managing stress in contemporary society. By consciously integrating digital free spaces into daily life, individuals can mitigate the adverse effects of constant connectivity, promoting mental well-being and a healthier balance between digital and offline activities.

8.6.2. Setting a Specific Period for Technology Use

Setting aside a specific period for technology use is a crucial strategy in managing stress through boundary setting. As the digital landscape continues to expand, the omnipresence of technology has significantly altered the dynamics of personal and professional lives. By delineating specific periods for technology engagement, individuals can mitigate stress and improve overall well-being. This discussion elucidates the importance of temporal boundaries in technology use.

Constant use of technology causes stress and cognitive strain, particularly in work environments. The idea that setting aside specific times for technology use might ease mental strain and improve cognitive performance is supported by this temporal boundary. People can avoid work intruding into their personal time by setting aside defined hours for checking emails or using digital devices, which can improve mental health and lower stress levels (Chang et al., 2015).

Turel, Serenko, & Giles's (2017) longitudinal study, which examines the impact of digital addiction on stress and well-being, highlights this issue even more. The authors found a link between higher levels of stress, anxiety, and depression with excessive technology use. They contend that establishing designated periods for technology use can aid in ending the vicious cycle of internet addiction. This tactic reduces impulsive consumption and encourages a balanced lifestyle by enabling controlled interaction with digital devices. According to Turel et al. (2017), temporal boundaries can be used as a useful intervention to reduce stress and enhance psychological health by reducing the obsessive demand to stay connected all the time.

Moreover, technology use and sleep disturbances are common stress-related issues. The researchers found that unrestricted use of technology, particularly before bedtime, significantly disrupts sleep patterns, leading to increased stress and fatigue. As a result, it is recommended to establish specific periods for technology use, especially by avoiding digital devices during the hour before sleep. This approach not only enhances sleep quality but also contributes to overall stress reduction. Chang et al. (2015) emphasize that temporal boundaries can help individuals create a more structured and health-conscious routine, thereby mitigating the adverse effects of technology on sleep and stress levels.

The synthesis of these studies emphasizes how crucial it is to designate particular times for using technology in order to reduce stress. Setting up such boundaries enables people to take charge of their digital use and keeps technology's all-pervasive influence from taking over daily life. By setting aside specific times for using technology, people can establish a healthy pattern that lowers stress and improves mental health. Therefore, the establishment of temporal boundaries for technology use is a vital strategy for stress management.

8.7. BALANCING TECHNOLOGY USE FOR MENTAL HEALTH

Managing technology's dual functions as a possible source of stress and a resource for assistance is necessary to strike a balance between its use and mental health. Understanding how digital platforms affect psychological well-being—from encouraging connectedness to inciting anxiety—is necessary to navigate this dynamic. Managing screen time strategically, utilizing digital tools for mindfulness and therapy, and addressing the concerns of excessive digital usage are all necessary to strike a healthy balance.

8.7.1. Evaluating Personal Technological Habits

A balanced approach to technology use that supports mental health must include an evaluation of one's technological habits. Academics have emphasized the need for self-evaluation and consciousness in controlling internet usage to reduce any negative consequences. Twenge et al. (2018) found that among teenagers and young adults, excessive screen time—especially on social media—correlates with higher levels of anxiety and sadness. This emphasizes how important it is for people to consider how they engage with technology to proactively manage mental health concerns (Twenge et al., 2018).

Moreover, a study by Lin et al. (2020) emphasizes the role of mindfulness in moderating technology's impact on psychological well-being. Mindful usage involves conscious awareness of one's digital interactions, fostering a more intentional and less compulsive approach to technology use. By practicing mindfulness, individuals can enhance their self-regulation abilities, leading to better management of stress and anxiety induced by technology (De Witte et al., 2021).

Additionally, a study conducted in 2018 by Economides et al. examines the significance of parental supervision and guidance in forming their children's digital habits. According to the study, parents who supervise their children's screen time and provide guidance when using technology can have a good impact on their mental health. Healthy technology habits can be fostered by establishing boundaries and encouraging balanced digital diets from a young age, which can help reduce the hazards associated with excessive screen time.

The key tactic for balancing technology use for mental health is to assess one's technological habits. It entails self-examination of one's online conduct, mindfulness in interaction techniques, and proactive parental supervision. Empirical evidence highlighting the relationship between digital consumption habits and mental health outcomes supports these strategies. Through the integration of these concepts, people can develop more positive interactions with technology, thus increasing mental well-being in the context of the ubiquitous presence of digital devices.

8.7.2. Recognizing Stressors Associated with Digital Use

Recognizing stressors associated with digital use is imperative for achieving a balanced approach to technology use that supports mental health. Several scholarly articles from recent years provide insights into this complex issue.

The role of excessive social media use can be a significant stressor. Constant connectivity and the pressure to maintain online personas can lead to anxiety and decreased well-being. This stressor is compounded by the phenomenon of "social comparison," where individuals compare their lives unfavorably to others based on curated online content, contributing to feelings of inadequacy and stress (Stanković et al., 2021).

Second, a study conducted in 2017 by Kuss and Griffiths examines the idea of "technological addiction" and how it affects mental health. They list several potential stressors that can interfere with daily schedules, sleep patterns, and interpersonal interactions, including excessive internet use, gaming, and smartphone dependency. These actions exacerbate stress and anxiety levels because they frequently result from underlying psychological needs like escapism or compulsive behavior. Moreover, the psychological processes that underlie improper usage of digital devices. Impulsivity, poor self-control, and the desire for rapid pleasure are some of the factors that lead to the emergence of digital addictions, which in turn raise stress levels and jeopardize general well-being.

It is critical for both individuals and mental health providers to identify these pressures. It makes it possible to put methods into place that lessen their effects and promote a positive connection with technology. According to Twidge (2019), interventions could involve supporting digital mindfulness practices, setting screen time limits, and encouraging offline activities that foster mental and emotional resilience.

A balanced approach to technology that promotes mental health must first identify and manage the stressors related to digital use. We can create well-informed strategies to manage the challenges of digital engagement while preserving psychological well-being by utilizing the insights from recent research.

8.7.3. Self-Regulation and Moderation Measures in Using Technology

Self-regulation refers to a person's capacity to observe, manage, and modify their behavior in response to external cues, such as those related to technology use. According to Maier et al. (2015), it includes techniques including limiting screen time, controlling digital distractions, and developing awareness of one's emotional and cognitive reactions to technology. Maier et al. (2015) claim that self-regulation encourages a deliberate approach to digital consumption, lowering the possibility of stress brought on by technology and enhancing psychological health.

Moderation measures further complement self-regulation by advocating balanced usage patterns that align with individual needs and contexts. The role of structured digital detoxes and designated offline periods in restoring cognitive resilience and enhancing overall mental health. These measures help counteract the potential overstimulation and cognitive fatigue associated with prolonged digital exposure.

Developing digital competence, or the ability to comprehend and use technology well, is also essential to encouraging balanced usage. Improving one's digital literacy enables people to use digital tools for stress relief and mental health assistance, as well as to safely traverse online environments and identify reliable information. This proficiency fosters proactive engagement with technology that is in line with personal well-being objectives in addition to improving self-regulation abilities.

Self-regulation and moderation measures constitute essential pillars in the pursuit of balanced technological use for mental health. They empower individuals to harness technology's benefits while mitigating its potential risks, thereby promoting resilience against digital stressors and fostering sustainable digital habits conducive to psychological well-being.

CHAPTER 9

STRESS AND SLEEP

Contents

The intricate relationship between stress and sleep is a pivotal area of investigation within the fields of psychology and neuroscience. Sleep, a fundamental biological process, is crucial for maintaining cognitive function, emotional regulation, and overall physical health. Conversely, stress, characterized by the body's response to perceived threats or challenges, can significantly impact sleep patterns and quality. Research indicates a bidirectional relationship wherein stress can lead to sleep disturbances, and insufficient sleep can exacerbate stress levels (Geiker et al., 2018) (Figure 9.1).

Figure 9.1. *The cycle of stress and sleeplessness.*

Source: Cordell (2024).

This link is underpinned by intricate and diverse pathophysiological mechanisms. The glucocorticoid hormone cortisol, which is released by the hypothalamic-pituitary-adrenal (HPA) axis, is crucial in the stress response because it helps the body get ready for a "fight or flight" reaction. High cortisol levels have been linked to trouble falling and staying asleep, especially in the evening (Kalmbach et al., 2018). According to Geiker et al. (2018), stress-induced hyperarousal, which is defined by elevated sympathetic nervous system activity, can also interfere with the sleep cycle's homeostatic regulation, resulting in insomnia and other sleep disorders.

Chronic stress is a strong predictor of poor sleep quality, as empirical research has shown. A strong association has been shown between elevated stress levels and an increase in sleep disturbances using the Pittsburgh Sleep Quality Index (PSQI), example, which has been widely used to measure the effect of stress on sleep (Geiker et al., 2018). Longitudinal research also indicates that people with high stress are more likely to develop chronic insomnia, which can have detrimental implications on mental health, including an increased vulnerability to depression and anxiety (Kalmbach et al., 2018).

Going forward, the chapter will explore various aspects of the stress-sleep nexus including strategies for maintaining sleep hygiene to improve overall health.

9.1. EFFECTS OF STRESS ON SLEEP

Chronic stress exerts profound effects on sleep, leading to a cascade of detrimental outcomes. The key effects of chronic stress on sleep, as evidenced by academic research are discussed in subsections.

9.1.1. Insomnia

Chronic stress precipitates insomnia through a complex interplay of neurobiological and psychological mechanisms. Central to this process is the dysregulation of the HPA axis, which plays a pivotal role in the body's stress response. Under chronic stress, the HPA axis remains persistently activated, resulting in elevated levels of cortisol, particularly during evening hours. Elevated cortisol disrupts the natural circadian rhythm, impeding the ability to initiate and maintain sleep (Geiker et al., 2018).

A state of hyperarousal, which is marked by elevated sympathetic nervous system activity and elevated cognitive and emotional arousal, is also brought on by prolonged stress. The symptoms of this hyperarousal state include a racing heartbeat, raised blood pressure, and intrusive thoughts, all of which prevent the relaxation required for the onset of sleep (Kalmbach et al., 2018).

Psychologically, long-term stress leads to maladaptive thought patterns and actions, such as worrying and rumination, which make it harder to fall asleep. A vicious cycle of insomnia is perpetuated when people experiencing chronic stress engage in negative cognitive appraisal of sleep, which increases worry about sleep itself (Kalmbach et al., 2018).

According to Geiker et al. (2018), empirical research continuously shows a significant link between long-term stress and insomnia, highlighting the significance of treating stress with therapeutic interventions in order to return to regular sleep patterns (Figure 9.2).

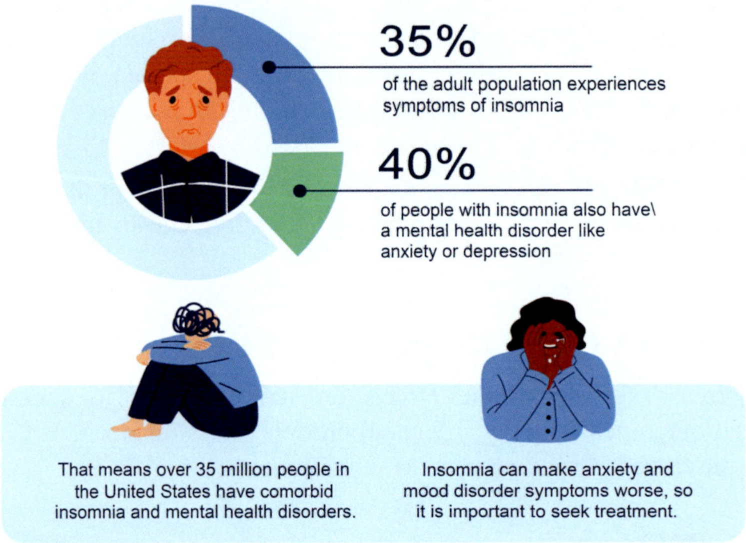

Figure 9.2. *Insomnia worsens mood, anxiety, and suicide risk.*

Source: Suni & Dimitriu (2024).

9.1.2. Sleep Fragmentation

Sleep fragmentation is caused by chronic stress via both psychological and physiological mechanisms. Chronic stress physiologically promotes the HPA axis's persistent activation, which raises cortisol levels and prolongs sympathetic nervous system activity. The natural sleep-wake cycle is disturbed by this hyperarousal, which also encourages numerous awakenings during the night (Kalmbach et al., 2018). High cortisol levels make it difficult to get a full night's sleep, especially during the nocturnal phase (Geiker et al., 2018). This results in fragmented sleep patterns.

Psychologically, long-term stress causes elevated emotional and cognitive arousal, which is manifested as excessive anxiety, ruminating, and intrusive thoughts. These psychological conditions may result in recurrent awakenings and trouble falling back asleep after waking up (Kalmbach et al., 2018). Multiple nocturnal awakenings interrupt the continuity and quality of sleep, which is a manifestation of the interplay between psychological stress and physiological hyperarousal (Yaribeygi et al., 2017).

Empirical research underscores the significant impact of chronic stress on sleep fragmentation, highlighting the need for interventions that address both the physiological and psychological aspects of stress to improve sleep continuity and overall sleep quality (Kalmbach et al., 2018). Effective management of chronic stress through cognitive-behavioral therapy, mindfulness practices, and stress-reduction techniques can mitigate its adverse effects on sleep (Khoury et al., 2015).

9.1.3. Altered Sleep Architecture

Chronic stress profoundly alters sleep architecture by disrupting the balance and duration of sleep stages. This disturbance is primarily mediated through the persistent activation of the HPA axis, resulting in elevated cortisol levels and sustained sympathetic nervous system activity (Scott et al., 2021). Elevated cortisol interferes with the normal progression of sleep stages, reducing the amount of slow-wave sleep (SWS) and rapid eye movement (REM) sleep, both of which are critical for restorative sleep and cognitive function (Bonnar et al., 2018).

Furthermore, long-term stress creates a condition known as hyperarousal, which is characterized by a rise in the frequency and severity of nighttime awakenings. This disrupts sleep even further and modifies the circadian rhythm (Bonnar et al., 2018). The body's capacity to heal and perform at its best is hampered by this disturbance, which lowers the quantity and quality of sleep.

9.1.4. Increased Sleep Latency

Chronic stress leads to increased sleep latency, the time it takes to transition from wakefulness to sleep, through both physiological and psychological mechanisms. Physiologically, chronic stress activates the hypothalamic-pituitary-adrenal (HPA) axis, resulting in elevated cortisol levels and heightened sympathetic nervous system activity. This state of hyperarousal makes it difficult for the body to relax and initiate sleep (Kalmbach et al., 2018).

Chronic stress psychologically encourages worry, rumination, and intrusive thoughts, all of which make it harder to fall asleep. Chronic pre-sleep arousal, which is characterized by an inability for people to disengage from stressful thoughts and quiet their minds, can result from the cognitive and emotional arousal brought on by stress (Geiker et al., 2018). People who engage in this pre-sleep cognitive activity are more likely to experience greater sleep delay because they can't easily go from wakefulness to sleep (Scott et al., 2021).

9.1.5. Daytime Sleepiness and Fatigue

Chronic stress mostly affects the quantity and quality of sleep at night, which causes weariness and drowsiness during the day. High cortisol levels and increased sympathetic nervous system activity are caused by the HPA axis's persistent activation in response to chronic stress, which leads to decreased slow-wave and REM sleep as well as fragmented sleep (Kalmbach et al., 2018). Non-restorative sleep experiences result from these disturbances to sleep's restorative capacities.

Additionally, chronic stress induces a state of hyperarousal and cognitive intrusion, characterized by worry and rumination, which further fragment sleep and extend sleep latency (Scott et al., 2021). The cumulative effect of poor sleep quality manifests as excessive daytime sleepiness and fatigue, impairing cognitive performance, mood, and overall daytime functioning (Bonnar et al., 2018). Thus, managing chronic stress is critical to improving sleep quality and mitigating daytime impairments.

9.2. STRATEGIES FOR IMPROVING SLEEP QUALITY

Improving sleep quality is critical for overall health and well-being. The following strategies, supported by academic research, are effective in enhancing sleep quality:

9.2.1. Regular Physical Activity

Regular physical activity is widely recognized as an effective strategy for improving sleep quality, with a substantial body of research elucidating its multifaceted benefits. Engaging in regular exercise helps regulate the circadian rhythm, the body's internal clock that governs sleep-wake cycles, thereby promoting more consistent and restorative sleep patterns (Scott et al., 2021).

Physical activity influences sleep architecture by increasing the duration of slow-wave sleep (SWS), the deep, restorative phase of sleep crucial for physical and mental recovery (Bonnar et al., 2018). Exercise also facilitates the reduction of sleep onset latency, the time it takes to transition from wakefulness to sleep, by promoting physical fatigue and mental relaxation (Geiker et al., 2018). This reduction is particularly beneficial for individuals suffering from insomnia, as it helps counteract the prolonged periods of wakefulness often associated with this condition.

Additionally, regular exercise reduces thc negative effects of stress, which is known to interfere with sleep. Exercise causes physiological changes that improve mood and reduce stress, including the release of endorphins and a decrease in cortisol levels (Scott et al., 2021). Lower stress levels are linked to fewer nocturnal awakenings and more consolidated sleep, which makes stress reduction essential for enhancing the quality of sleep. Research based on empirical data supports the beneficial effects of exercise on sleep. According to a meta-analysis by Scott et al. (2021), regular exercise dramatically increases the effectiveness and quality of sleep across a variety of demographics. Another study by Kalmbach et al. (2018) showed that in those with persistent primary insomnia, moderate-intensity aerobic exercise reduced symptoms of insomnia and enhanced the quality of sleep.

9.2.2. Consistent Sleep Schedule

Maintaining a consistent sleeping pattern is essential for enhancing the quality of your sleep because it helps control the circadian rhythm, the body's internal clock that cycles between sleep and wakefulness. Regular wake and sleep schedules help to synchronize circadian rhythms, which in turn promotes more restorative and consistent sleep (Bonnar et al., 2018). This natural cycle is disturbed by irregular sleep schedules, which makes it harder to initiate and maintain sleep and lowers the quality of sleep (Scott et al., 2021).

Adherence to a consistent sleep schedule stabilizes the timing of key physiological processes, including the release of melatonin, a hormone that signals the onset of sleep (Kalmbach et al., 2018). Consistency in sleep timing also enhances sleep efficiency and reduces sleep onset latency, facilitating a quicker transition into sleep and reducing nocturnal awakenings. Empirical evidence highlights the positive correlation between regular sleep patterns and improved sleep quality, underscoring the importance of maintaining consistent sleep routines for optimal sleep health (Geiker et al., 2018).

9.2.3. Mindfulness-Based Stress Reduction

The main ways that Mindfulness-Based Stress Reduction (MBSR) works to improve the quality of sleep are by reducing stress and encouraging relaxation. MBSR incorporates techniques like mindfulness meditation, which lower cognitive arousal and foster present-moment awareness. This decrease in cognitive arousal lessens the intrusive thoughts and rumination that frequently hinder the start and maintenance of sleep (Perrier et al., 2020).

According to Khoury et al. (2015), MBSR physiologically reduces the activation of the HPA axis, which lowers cortisol levels and lowers sympathetic nervous system activity, both of which promote better sleep. Empirical research confirms that MBSR improves the quality of sleep. For example, Goyal et al. (2014)'s meta-analysis showed that MBSR dramatically enhances sleep metrics, such as total sleep time and sleep efficiency, in those with sleep disorders.

Since it promotes a mental and physical state that is conducive to restorative sleep, MBSR is an effective intervention for enhancing the quality of sleep.

9.2.4. Avoiding Exposure to Electronics Before Bedtime

Limiting exposure to electronic devices at bedtime is a strategic approach to improving sleep quality by minimizing blue light interference with melatonin production. Electronic devices, such as smartphones, tablets, and computers, emit blue light that can suppress melatonin, a hormone crucial for regulating sleep-wake cycles (Chang et al., 2015). Melatonin suppression delays sleep onset and disrupts the circadian rhythm, leading to poor sleep quality and reduced sleep duration (Geiker et al., 2018)

Empirical studies back up the negative impacts of using electronics right before bed on sleep. For example, Chang et al. (2015) discovered that using electronics more before bed was linked to higher levels of daytime sleepiness, worse sleep efficiency, and longer sleep latency. Individuals can maintain higher levels of melatonin, which will promote more restorative sleep and a smoother transition to sleep by minimizing their screen time and exposure to blue light before bedtime (Stanković et al., 2021).

9.3. CREATING A RELAXING BEDTIME ROUTINE

Creating a relaxing bedtime routine is essential for enhancing sleep quality and ensuring restorative sleep. The following five strategies, supported by academic research, are effective in establishing a calming pre-sleep environment:

9.3.1. Engaging in Relaxation Techniques

Engaging in relaxation techniques is an effective strategy for creating a relaxing bedtime routine by reducing physiological and psychological arousal, thereby promoting a state conducive to sleep. Techniques such as progressive muscle relaxation, deep breathing exercises, and guided imagery help activate the parasympathetic nervous system, which counteracts the stress-induced activation of the sympathetic nervous system (Stanković et al., 2021). This physiological shift leads to decreased heart rate, lower blood pressure, and reduced cortisol levels, facilitating the transition to sleep (Morales-Muñoz & Gregory, 2023).

Additionally, relaxation techniques mitigate cognitive arousal by diverting attention from stressors and intrusive thoughts, which are common barriers to sleep onset (Baranwal et al., 2023). Empirical studies have demonstrated that regular practice of relaxation exercises before bedtime significantly improves sleep quality, reduces sleep onset latency, and enhances overall sleep efficiency (Baranwal et al., 2023). Thus, incorporating relaxation techniques into a bedtime routine is a scientifically supported approach to achieving better sleep.

9.3.2. Reading a Book

A good way to establish a calming nighttime routine is to read a book before bed. This practice helps to promote cognitive relaxation and redirect attention from stressors. Lowering mental arousal and preparing the mind for sleep can be achieved by partaking in a low-stimulation activity, like reading a physical book (Morales-Muñoz & Gregory, 2023). Pre-sleep anxiety and ruminative thoughts, which frequently prevent sleep from starting, can be reduced with this cognitive diversion technique (Bonnar et al., 2018).

Additionally, reading under dim lighting conditions, as opposed to the blue light emitted by electronic devices, minimizes disruption to the production of melatonin, a hormone critical for sleep regulation (Morales-Muñoz & Gregory, 2023). By establishing a consistent pre-sleep reading routine, individuals can condition their bodies and minds to associate this activity with the onset of sleep, thus enhancing overall sleep quality and efficiency (Geiker et al., 2018).

9.3.3. Mindfulness and Meditation

Practicing mindfulness and meditation is an effective strategy for creating a relaxing bedtime routine by fostering mental calmness and reducing pre-sleep cognitive arousal. Mindfulness meditation involves focusing on the present moment and cultivating awareness without judgment, which helps to quiet the mind and alleviate anxiety and stress that often hinder sleep onset (Bonnar et al., 2018). This practice decreases the activity of the sympathetic nervous system and reduces cortisol levels, promoting physiological relaxation conducive to sleep.

Empirical studies substantiate the effectiveness of mindfulness meditation in enhancing the caliber of sleep. According to a study by Baranwal et al. (2023), people who practiced mindfulness meditation saw a significant decrease in sleep disruptions and an improvement in sleep efficiency. People who incorporate mindfulness and meditation into their evening routine can establish a regular and peaceful sleep-inducing ritual that improves both the quality and length of their sleep.

9.3.4. Taking a Bath

Warm baths before bed help to enhance thermoregulatory processes that lead to sleep, which is a good way to establish a calming nighttime ritual. Warm water immersion raises body temperature, and when you get out of the bath, the quick cooling effect causes your core body temperature to drop, simulating the thermoregulatory drop that happens when you go to sleep (Morales-Muñoz & Gregory, 2023). This physiological shift improves sleep initiation and quality by alerting the body to get ready for sleep.

Additionally, taking a warm bath can help release physical tension and relax muscles, which promotes calmness and lessens the chance of awakening during the night (Baranwal et al., 2023). This practice is supported by empirical facts, as studies show that taking a bath before bed enhances both the overall quality and efficiency of sleep. Individuals who incorporate a warm bath into their evening routine can establish a calming and regular pre-sleep ritual that promotes restorative sleep.

9.3.5. Listening to Soothing Music

Listening to soothing music before bedtime is an effective strategy for creating a relaxing bedtime routine by promoting physiological and psychological relaxation. Music with a slow tempo and low volume can decrease heart rate, respiratory rate, and blood pressure, thereby activating the parasympathetic nervous system, which is conducive to sleep (Scott et al., 2021). This physiological relaxation helps mitigate stress and anxiety, common impediments to sleep onset.

The positive effects of music on sleep quality are supported by empirical research. For example, Baranwal et al.'s study from 2023 indicated that those with sleep difficulties who listened to relaxing music experienced a considerable improvement in their sleep efficiency and quality. The power of music to reduce outside noise and create a relaxing auditory environment contributes to the peacefulness of the pre-sleep atmosphere. Individuals can establish a regular and peaceful environment that improves the quality of their sleep by including relaxing music in their nighttime routine.

9.4. ADDRESSING SLEEP DISORDERS CAUSED BY STRESS

Stress is a significant etiological factor in various sleep disorders, with insomnia being the most prevalent. Insomnia, characterized by difficulties in sleep initiation, maintenance, or early morning awakenings, is often precipitated and exacerbated by chronic stress (Baranwal et al., 2023). Stress-induced hyperarousal, both physiological and cognitive, interferes with the sleep-wake cycle, leading to persistent sleep disturbances. Individuals experiencing high levels of stress may exhibit heightened sympathetic nervous system activity and elevated cortisol levels, which disrupt the natural sleep process and contribute to insomnia (Scott et al., 2021).

Another common sleep disorder associated with stress is sleep apnea, particularly obstructive sleep apnea (OSA). Chronic stress can exacerbate the severity of OSA by increasing the risk of airway collapse during sleep due to heightened muscle tension and inflammatory responses (Turner et al., 2020).

Additionally, stress-induced weight gain and metabolic dysregulation can contribute to the development and worsening of OSA (Geiker et al., 2018). The interplay between stress and OSA creates a detrimental cycle, as sleep fragmentation and hypoxia further elevate stress levels, leading to a perpetuation of both conditions.

Addressing stress through behavioral and psychological interventions is thus crucial in the management and mitigation of these stress-related sleep disorders. The following strategies, supported by academic research, are effective in mitigating the adverse effects of stress on sleep:

9.4.1. Cognitive Behavioral Therapy for Insomnia

Cognitive Behavioral Therapy for Insomnia (CBT-I) is a highly effective strategy for addressing sleep disorders arising from chronic stress. CBT-I combines cognitive and behavioral techniques to target the dysfunctional thoughts and behaviors that perpetuate insomnia. Key components include cognitive restructuring to challenge maladaptive beliefs about sleep, sleep restriction therapy to consolidate sleep, and stimulus control to reinforce the bed as a cue for sleep rather than wakefulness (Kalmbach et al., 2018).

There is empirical data to support CBT-I's effectiveness in enhancing sleep outcomes. According to Morales-Muñoz & Gregory (2023), CBT-I dramatically improves sleep quality, lowers sleep onset latency, and reduces wakefulness after sleep begins. CBT-I targets the cognitive and physiological elements of stress-related insomnia by reducing stress-induced hyperarousal and anxiety (Kalmbach et al., 2018). CBT-I, a non-pharmacological intervention, offers long-lasting and sustainable benefits to those with chronic stress and related sleep disorders.

9.4.2. Mindfulness-Based Stress Reduction

MBSR is a useful technique for treating chronic stress-related sleep problems. According to Geiger et al. (2016), mindfulness meditation techniques that improve present-moment awareness and encourage relaxation are part of MBSR, which lowers the physiological and psychological reactivity linked to stress. MBSR reduces stress and anxiety, which are significant causes of sleep disturbances, by concentrating attention on the present and developing a non-judgmental attitude toward thoughts and emotions (Kalmbach et al., 2018).

The advantages of MBSR for enhancing the quality of sleep are supported by empirical data. Sharma & Rush (2014) conducted a systematic study and discovered that MBSR considerably lowers sleep latency and increases sleep quality in people who suffer from insomnia. Furthermore, it has been demonstrated that MBSR lowers cortisol levels and sympathetic nervous system activity, promoting a sleep-promoting state (Khoury et al., 2015). Hence, MBSR is a useful intervention for stress-related sleep disturbances because it tackles both the cognitive and physiological aspects of stress.

9.4.3. Exercise and Physical Activity

Exercise and physical activity are effective strategies for mitigating sleep disorders stemming from chronic stress by promoting both physiological and psychological well-being. Regular physical activity has been shown to reduce symptoms of anxiety and depression, which are common contributors to sleep disturbances (Sharon-David & Tenenbaum, 2017). Exercise enhances sleep quality by regulating circadian rhythms and reducing the time taken to fall asleep, known as sleep latency (Bonnar et al., 2018).

Particularly, aerobic exercise has been shown to increase slow-wave sleep, the most restorative sleep stage, which increases sleep efficiency and length (Bonnar et al., 2018). Furthermore, by lowering cortisol levels and encouraging muscular relaxation, physical activity lessens the hyperarousal linked to stress. Research indicates that frequent participation in moderate-intensity exercise might improve sleep quality considerably and mitigate the negative impact of stress on sleep patterns (Sharon-David & Tenenbaum, 2017).

9.4.4. Pharmacotherapy

Pharmacotherapy serves as a pivotal strategy in managing sleep disorders induced by chronic stress, a prevalent issue affecting numerous individuals globally. Chronic stress disrupts normal sleep patterns by altering neurochemical balances and circadian rhythms, thereby necessitating targeted pharmacological interventions. Benzodiazepines, such as lorazepam and diazepam, act on GABA receptors to induce sedation and mitigate anxiety, which often accompanies chronic stress-related sleep disturbances (Kalmbach et al., 2018). Alternatively, non-benzodiazepine sedative-hypnotics like zolpidem and zaleplon provide sleep-promoting effects with reduced risk of dependence compared to traditional benzodiazepines (Turner et al., 2020).

Moreover, antidepressants like trazodone and mirtazapine, through their sedative properties and modulation of neurotransmitter systems, offer effective treatment for insomnia associated with chronic stress (Kalmbach et al., 2018). However, caution must be exercised due to potential side effects and interactions with existing conditions or medications. Consequently, pharmacotherapy tailored to individual needs and guided by clinical assessments remains a cornerstone in managing sleep disorders originating from chronic stress, offering relief and promoting overall well-being.

9.5. SEEKING PROFESSIONAL HELP FOR SLEEP ISSUES

Challenges with sleep are frequent concerns that have a big influence on general health and well-being. Stress is sometimes made worse by persistent sleep problems, creating a vicious cycle that harms both everyday functioning and mental health. Getting professional assistance to address this issue is essential to ending the cycle of stress-related sleep issues. Research indicates that treatments including medical consultations for underlying disorders and cognitive-behavioral therapy for insomnia (CBT-I) can effectively restore good sleep patterns (Perrier et al., 2020). The discussion below examines the need for professional assistance when dealing with sleep problems, emphasizing the advantages of customized therapy techniques in reducing stress and improving sleep quality.

9.5.1. Seeking for Professional Aid: Physician or Psychologist?

It's important to choose the right practitioner for treating insomnia. Physicians who specialize in primary care, sleep medicine, cognitive behavioral therapy (CBT), and psychiatry are all essential in identifying underlying mental health disorders (Perrier et al., 2020). A comprehensive strategy that addresses both symptoms and underlying causes is ensured via collaborative care.

There are several important factors to take into account while seeking professional help for sleep problems:

1. **Speak with Your Primary Care Physician:** Primary care physicians are qualified to provide early advice and referrals to other specialists who are skilled in treating sleep issues.
2. **Consult with Sleep Specialists:** These professionals are qualified to identify and manage disorders like insomnia.
3. **Examine Psychological Therapies:** Cognitive-behavioral therapy (CBT), which is especially intended for insomnia, can be administered by psychologists and therapists.
4. **Consider Considering Psychiatric Evaluation:** Psychiatrists are qualified to diagnose and treat any underlying mental health issues that may be causing sleep difficulties.
5. **Take a Collaborative Approach:** Including a multidisciplinary team makes individualized treatment plans and comprehensive care easier to implement.

- **Role in Diagnosing and Treating Insomnia:** Sleep specialists determine causes and contributing factors for insomnia by evaluating symptoms and medical histories. Customized treatment regimens, involving both pharmaceutical and non-pharmacological alternatives, are designed to improve general well-being and promote sound sleep.

- **Collaborative Method for Holistic Care:** For the holistic treatment of insomnia, a collaborative method is necessary. In order to ensure efficient and long-lasting management, healthcare practitioners collaborate to create thorough, individualized treatment programs that address the symptoms of insomnia as well as any underlying mental health conditions.

9.5.2. The Importance of Seeking Professional Help

9.5.2.1. Identifying Underlying Mental Health Issues

It's critical to understand that underlying mental health conditions might occasionally be connected to sleeplessness. This suggests that there might be additional causes, such as anxiety or dismay, for your inability to go to sleep.

For treatment to be effective, it is essential to identify these underlying problems. Your mental health can be evaluated by a medical practitioner to see whether it's contributing to your insomnia.

Recall that improving these underlying issues can improve your general well-being and quality of sleep.

9.5.2.2. Tailored Treatment Schedules

Since everyone's experience with insomnia is different, therapy regimens must be customized for each patient. You should anticipate that when you seek expert assistance for insomnia, medical specialists will develop individualized treatment programs based on your unique requirements and situation.

These plans evaluate things like the severity of your symptoms, possible underlying reasons, and any concurrent physical or mental health issues. Healthcare providers can boost your chances of recovering from insomnia by creating a treatment plan that is specially designed for you.

A common approach that makes up customized treatment programs for insomnia is cognitive behavioral therapy (CBT). The goal of this therapy is to alter unfavorable thought and behavior patterns that interfere with sleep. It assists by teaching you how to relax before bed, control your anxious feelings, create a consistent sleep schedule, and adopt healthy sleep hygiene practices like putting electronics away early. A customized treatment plan may also involve medications if needed (Perrier et al., 2020).

9.5.2.3. Long-Term Management

Long-term management of insomnia is essential for enhancing sleep quality and general well-being. Any underlying mental health conditions that might be causing your insomnia should be addressed.

By getting professional assistance, you can get a customized treatment plan that meets your needs. This could involve a mix of therapeutic approaches, such as cognitive behavioral therapy (CBT), and, if required, pharmaceutical choices.

It has been demonstrated that CBT is useful in enhancing sleep quality and assisting people in forming sound sleeping habits. It's never too late to get treatment for your insomnia, and with the correct assistance, you can control your condition long-term and get adequate rest every night.

9.5.2.4. Importance of Therapy for Sleep Disorders

Therapy, such as Cognitive Behavioral Therapy (CBT), is crucial for addressing sleep disorders by teaching effective techniques and strategies to improve sleep quality (Baranwal et al., 2023).

9.5.2.5. CBT for Insomnia and Other Sleep Disorders

If you're struggling with insomnia or other sleep disorders, one effective treatment option is Cognitive Behavioral Therapy (CBT). CBT focuses on changing negative thoughts and behaviors that contribute to sleep problems.

It's been shown to improve sleep patterns and quality for people with insomnia. The therapy includes techniques like psychoeducation, where you learn about healthy sleep habits and how they impact your overall well-being.

Sleep hygiene counseling is another aspect of CBT, which involves creating a bedtime routine and making changes to your environment for better sleep. Relaxation techniques are also taught in therapy, such as deep breathing exercises and progressive muscle relaxation (Baranwal et al., 2023).

9.5.3. How CBT Works?

The goal of cognitive behavioral therapy (CBT), an efficient treatment for insomnia, is to alter beliefs and actions that may contribute to sleep disturbances. CBT offers you new ways of thinking about sleep and assists you in identifying negative thought patterns.

They also assist you in creating plans to enhance your sleeping patterns. Stimulus control treatment is one method utilized in cognitive behavioral therapy. This entails establishing a regular bedtime routine and reserving the bed for sleeping or alone time. Another method is called sleep restriction therapy, and it entails setting a bedtime limit that corresponds with your real sleep duration.

CBT also includes relaxation techniques, such as deep breathing exercises or progressive muscle relaxation, to help calm the mind and body before bed. It may take some time and practice, but many people who try CBT find that it improves their sleep quality and helps them fall asleep faster.

9.5.4. Behavioral Techniques Used in Therapy

In therapy for insomnia, there are behavioral techniques that can help improve sleep. These techniques include (Figure 9.3):

1. Sleep Hygiene: Establishing a consistent sleep schedule and creating a relaxing bedtime routine.
2. Stimulus Control: Associating the bed with sleep by using it only for sleeping or intimate activities.
3. Relaxation Techniques: Practicing relaxation exercises such as deep breathing or progressive muscle relaxation before bedtime.
4. Cognitive Restructuring: Identifying and challenging negative thoughts or worries that may contribute to insomnia.
5. Biofeedback: Using instruments to monitor physical signs of tension and teaching techniques to control them.
6. Sleep Restriction: Limiting the time spent in bed to increase sleep drive and promote better sleep (Kalmbach et al., 2018).

Figure 9.3. *CBT techniques.*

Source: Wise (2022).

9.5.5. The Importance of a Multi-Disciplinary Approach

In order to provide complete care, a multidisciplinary approach to stress management integrates several professional views and methods. PCPs, psychologists, psychiatrists, and other medical experts work together in this strategy. Multidisciplinary approaches, such as medical treatment, psychological therapy, and lifestyle modifications, are more successful than single-modality treatments in controlling stress, according to a 2018 study by Kalmbach et al For example, addressing both the psychological and physiological elements of stress can be achieved by combining CBT with medicine prescribed by a psychiatrist. Treatment outcomes can also be improved by adding wellness programs including stress management workshops, dietary counseling, and physical activity. By addressing all aspects of a person's health, the multidisciplinary approach promotes holistic well-being.

9.6. IMPORTANCE OF ADEQUATE SLEEP FOR STRESS MANAGEMENT

For the best possible functioning of the brain, emotions, and body, sleep is a basic biological process. Getting enough sleep is essential for stress management since it affects the body's capacity to handle everyday stressors. The relationship between stress and sleep is bidirectional, as evidenced by scientific research: stress can worsen sleep patterns, while poor sleep can make stress worse. It follows that getting enough good sleep is essential for preserving emotional equilibrium, improving mental function, and fostering general well-being. In addition to highlighting the importance of giving sleep a top priority in stress management measures, this introduction seeks to clarify the mechanisms by which sleep modifies stress reactions.

9.6.1. Mechanism by Which Sleep Modifies Stress Response

The relationship between sleep and stress is deeply intertwined, with sleep serving as a critical modulator of stress responses. Understanding the mechanisms through which sleep influences stress reactions provides valuable insights into developing effective stress management strategies.

The influence of sleep on the hypothalamic-pituitary-adrenal (HPA) axis, a key element of the body's stress response system, is one of the main ways that sleep alters stress reactivity. One important stress hormone, cortisol, is secreted under the control of the HPA axis. The circadian pattern of cortisol production, which normally peaks in the early morning and decreases during the day, depends on getting enough sleep. Sleep disturbances, especially those brought on by insomnia or sleep loss, can cause the HPA axis to become dysregulated, which raises cortisol levels. The body's capacity to handle stress is hampered by this dysregulation, which intensifies stress reactions (Khoury et al., 2015).

Neuroimaging studies have provided further evidence of sleep's role in modulating stress reactions by examining brain regions involved in emotional regulation. The prefrontal cortex (PFC) and the amygdala are two such regions. The PFC is responsible for executive functions, including decision-making and emotional regulation, while the amygdala processes emotional responses, particularly fear and anxiety. Sleep deprivation has been shown to reduce PFC activity and increase amygdala reactivity, leading to heightened emotional responses and impaired emotional regulation (Geiker et al., 2018). Consequently, individuals who experience poor sleep are more likely to have exaggerated stress reactions and difficulty managing emotional stress.

Additionally, sleep makes it easier for emotional memories to stick together, which is important for handling stress. Sleep, and especially rapid eye movement (REM) sleep, allows the brain to integrate and process emotional experiences from the previous day. Consolidation facilitates adaptive coping mechanisms and lessens the emotional impact of traumatic memories (Geiker et al., 2018). This process can be hindered by REM sleep disruption, which can result in prolonged emotional suffering and increased stress responses.

Sleep affects stress reactions primarily because of its role in immunological function. It is well-recognized that long-term stress weakens the immune system and makes people more prone to sickness. On the other hand, getting enough sleep strengthens the immune system, in part because it increases the production of cytokines that help fight inflammation and infections (Stanković et al., 2021). Enough sleep supports immunological function, which lessens the damaging effects of stress on the body, lowering total stress levels and fostering resilience.

Furthermore, sleep affects mood and mental health, which in turn affects stress reactions. According to Baglioni et al. (2016), there is a clear correlation between the development of mood disorders such as anxiety and depression and chronic sleep deprivation. Increased susceptibility to stress and compromised coping strategies are hallmarks of these illnesses. People who get enough sleep are better able to sustain mental health and mood stability, which improves their capacity to handle stress.

Thus, sleep alters stress responses via a variety of interrelated processes, such as HPA axis control, emotional regulation-related brain area modulation, emotional memory consolidation, immune system support, mood maintenance, and mental health maintenance. Therefore, getting enough good sleep is essential for stress management.

9.6.2. Importance of Prioritizing Sleep in Stress Management

Sleep is essential to health and well-being because it helps us cope with stress. Despite its significance, sleep is frequently overlooked in today's culture of constant stress. The complex relationship that exists between stress and sleep emphasizes how important it is to prioritize sleep as a cornerstone of stress reduction techniques.

Firstly, sleep is critical to general health, emotional control, and cognitive performance. Stanković et al. (2021) assert that sleep is necessary for restorative, emotional, and memory consolidation processes—all of which are critical for efficient stress management. When we sleep, especially in the rapid eye movement (REM) phase, the brain organizes feelings and stores them in memory. This procedure improves emotional resilience by assisting people in comprehending and managing stressors more effectively. Conversely, inadequate sleep compromises these cognitive processes, resulting in increased stress reactivity and poor coping strategies.

The reciprocal link between stress and sleep emphasizes how important it is to prioritize getting enough sleep to interrupt the pattern of stress-related sleep disruptions. Cortisol and other stress hormones are released when stress triggers the hypothalamic-pituitary-adrenal (HPA) axis, which can disrupt sleep. Persistent hyperarousal brought on by chronic stress might make it challenging to fall asleep and stay asleep. On the other hand, lack of sleep makes stress reactions worse by raising the HPA axis' activity and lowering the threshold for stress (Bonnar et al., 2018). It is crucial to prioritize getting enough sleep to reduce stress because of this vicious cycle of interactions between stress and sleep, which makes it worse to sleep and increases stress.

Making sleep a priority can greatly improve both psychological and physiological stress resilience. A study by Kalmbach et al. (2016) found that people who prioritize sleep report feeling less stressed and are more resilient to stressors. This is partially due to the fact that getting enough sleep enhances metabolic health, lowers inflammation, and strengthens the immune system—all of which are factors in an individual's capacity to handle stress. Psychologically, getting enough sleep elevates mood, lowers anxiety, and increases general well-being, which makes people better able to manage stress.

In addition to the individual benefits, prioritizing sleep has broader implications for societal health. Sleep deprivation is linked to various adverse outcomes, including impaired academic and occupational performance, increased risk of accidents, and higher healthcare costs due to stress-related illnesses (Morales-Muñoz & Gregory, 2023). By fostering a culture that values sleep, society can mitigate these negative outcomes and promote better mental health and productivity. Employers, educational institutions, and policymakers should advocate for healthy sleep practices as part of comprehensive stress management programs.

Despite the evident advantages of sleep for stress reduction, obstacles still exist in reaching sufficient sleep. High work expectations, excessive screen time, and constant stress are characteristics of modern lifestyles that frequently compromise both the quantity and quality of sleep. It is crucial to implement interventions targeted at enhancing sleep hygiene, such as setting up regular sleep routines, making a sleep-friendly environment, and minimizing screen time before bed. Furthermore, it has been demonstrated that cognitive-behavioral treatment for insomnia (CBT-I) is beneficial in enhancing sleep and lowering stress, providing a workable option for people with persistent sleep difficulties (Morales-Muñoz & Gregory, 2023).

Considering sleep has such a profound impact on cognitive function, emotional control, and general health, it is essential for stress management. A proactive approach to stress management and sleep hygiene is required due to the mutually beneficial connection between stress and sleep. By stressing the value of sleep, people can become more resilient to stress, feel better overall, and help create a healthier society. Prioritizing sleep in stress management techniques is not only advantageous but also necessary to solve the widespread problem of stress in modern living.

9.7. BUILDING A HEALTHY SLEEP ROUTINE

Considering it promotes both physical and emotional well-being, developing a regular sleep habit is essential to controlling stress. A basic biological requirement, sleep has a major impact on cerebral processes, emotional control, and general wellness. Inadequate or disrupted sleep patterns increase stress, hinder judgment, and reduce resistance to day-to-day obstacles. People can improve their capacity to handle stress by making regular sleep routines a priority. Using contemporary academic research to highlight the significance of sleep hygiene practices in fostering good mental health and general life quality, this introduction explores the crucial role that scheduled sleep routines play in stress management.

9.7.1. ENGAGING IN DAILY WELLNESS ACTIVITIES

Daily wellness activities are essential for maintaining a sound sleep schedule, which in turn improves general health and reduces stress. Incorporating regular wellness activities improves sleep quality and supports the physiological and psychological balance necessary for restorative sleep, according to a study by Turner et al. (2020).

Maintaining regular routines that promote healthy sleep-wake cycles throughout the day is essential to creating a sound sleep schedule. Frequent exercise, for example, has been proven to enhance sleep onset latency and regulate circadian rhythms, supporting the proper coordination of biological processes essential for sleep (Buysse, 2014). Moderate exercise throughout the day can encourage deeper and more peaceful sleep at night, lowering the risk of stress-related sleep disruptions (Morales-Muñoz & Gregory, 2023).

Furthermore, dietary choices play a crucial role in sleep hygiene. Consuming a balanced diet rich in nutrients that support sleep, such as magnesium and tryptophan, can enhance sleep quality by facilitating the production of sleep-inducing neurotransmitters (Buysse, 2014). Conversely, excessive caffeine intake or consuming heavy meals close to bedtime may disrupt sleep patterns and exacerbate stress-related insomnia (Buysse, 2014).

Furthermore, a peaceful nighttime routine that encourages relaxation and facilitates the transition to sleep is enhanced by mindfulness exercises and stress-reduction strategies (Bonnar et al., 2018). By incorporating these wellness activities into their everyday routines, people can create an atmosphere that supports the development of regular, healing sleep patterns. In the end, a comprehensive approach to everyday wellness promotes mental toughness and physical health in addition to a regular sleep schedule that is necessary for stress reduction and general well-being.

9.7.2. Developing a Consistent Sleeping Routine

A regular sleep schedule is essential for promoting a healthy sleep pattern, which in turn greatly improves general well-being and reduces stress. Grandner & Hale (2016) found that in order to optimize sleep quality and duration—two key elements of a good sleep routine—regular sleep-wake cycles must be maintained.

A structured sleep routine involves going to bed and waking up at the same times each day, including weekends, to regulate the body's internal clock, known as the circadian rhythm (Bonnar et al., 2018). This consistency reinforces the body's natural sleep-wake cycle, promoting efficient transitions between wakefulness and sleep phases. Moreover, adhering to a regular bedtime routine signals the brain and body to anticipate sleep, facilitating faster onset and improved sleep efficiency (Morales-Muñoz & Gregory, 2023).

A regular sleep schedule also creates behavioral cues that get the body and mind ready for sleep. Reduced arousal and increased relaxation are crucial for both starting and sustaining sleep. Rituals including practicing relaxation techniques, turning down lights, and avoiding stimulating activities before bedtime help achieve this (Bonnar et al., 2018).

Importantly, the advantages of maintaining a regular sleep schedule go beyond just getting enough sleep. According to research, those with consistent sleep schedules have higher overall sleep quality, less daytime drowsiness, and enhanced cognitive function than people with erratic sleep schedules (Morales-Muñoz & Gregory, 2023). A regular sleep schedule improves both the quantity and quality of sleep, which strengthens stress resistance, improves mood control, and promotes healthy psychological performance.

Empirical evidence underscores the critical role of maintaining a regular sleep routine in cultivating a healthy sleep pattern. By aligning with natural circadian rhythms and fostering sleep-promoting behaviors, individuals can achieve better sleep outcomes, thereby contributing to overall health and stress management.

9.7.3. Avoid Alcohol and Caffeine

Sustaining a healthy sleep schedule requires avoiding alcohol and caffeine, which has a major negative influence on both sleep quality and general well-being. Both drugs have been shown to disturb sleep-wake cycles and impair sleep architecture, which can have an impact on one's physical and mental well-being (Buysse, 2014).

Alcohol, which is frequently taken as a relaxant, makes you feel sleepy at first but eventually breaks up your sleep. It modifies the activity of neurotransmitters, specifically gamma-aminobutyric acid (GABA), which can cause sleep patterns to become disjointed and the quality of sleep to decrease (Roehrs & Roth, 2018). Furthermore, alcohol worsens the restorative qualities of sleep by raising the prevalence of snoring and sleep apnea (Turner et al., 2020). Similarly, caffeine, a widely consumed stimulant, exerts its effects by antagonizing adenosine receptors, which regulate wakefulness. Its consumption close to bedtime delays the onset of sleep and reduces total sleep time, exacerbating sleep fragmentation (Bonnar et al., 2018). Chronic caffeine use can lead to tolerance, requiring higher doses to achieve the same wake-promoting effects, thereby perpetuating a cycle of sleep disruption (Turner et al., 2020).

The amount of caffeine and alcohol consumption can be reduced or eliminated to encourage a more regular sleep schedule and improve the quality of sleep. Kalmbach et al. (2018) report that this technique enhances the efficiency of sleep stages necessary for both physical restoration and cognitive performance, while also supporting the alignment of circadian rhythms. Following these behavioral changes, then, greatly aids in creating a regular sleep schedule that promotes stress reduction and general well-being.

CHAPTER 10

STRESS AND RESILIENCE

Contents

Stress and resilience are two intertwined constructs pivotal in understanding human adaptation to adversity. Stress, often conceptualized as a psychological and physiological response to challenging or threatening situations, is a ubiquitous aspect of the human experience. It encompasses a range of reactions, from acute stress responses, characterized by immediate, short-term reactions to specific stressors, to chronic stress, which involves prolonged exposure to stress-inducing conditions. The seminal work of Hans Selye in the mid-20th century introduced the notion of the General Adaptation Syndrome (GAS), delineating the stages of alarm, resistance, and exhaustion in response to sustained stress (Southwick et al., 2014).

Conversely, resilience is the capacity to deal with stress and adversity in a constructive manner. It is not just the absence of negative consequences when faced with stress; rather, resilience is a dynamic process of positive adaptation (Macedo et al., 2014). Resilience can be viewed from a variety of angles, including individual characteristics like optimism, self-efficacy, and cognitive flexibility, as well as external factors like social support, community resources, and environmental circumstances (Stanley, 2014). Crucially, resilience is not a fixed trait but rather a developmental process that can be fostered through focused interventions and supportive environments.

Resilience and stress have a complicated and context-dependent relationship. According to research, resilience can improve general functioning and well-being by reducing the harmful impacts of stress (Southwick et al., 2014). The significance of resilience-building techniques in stress management interventions is highlighted by this reciprocal link. By encouraging adaptive coping strategies and lowering stress-related symptoms, for example, mindfulness-based stress reduction (MBSR) programs have been demonstrated to improve resilience (Macedo et al., 2014).

Therefore, the study of stress and resilience is critical for developing effective strategies to enhance mental health and well-being. By fostering resilience, individuals can better withstand and recover from the inevitable stresses of life, leading to improved health outcomes and quality of life.

Going forward, this chapter will explore various resilience-building strategies that foster optimism for effectively navigating through stressful situations.

10.1. BUILDING RESILIENCE TO STRESS

Building resilience to stress involves various strategies that enhance an individual's capacity to cope with adversity and recover from stressful experiences. Scholarly sources highlight several evidence-based approaches to fostering resilience:

10.1.1. Mindfulness and Meditation

Through the promotion of improved emotional regulation and cognitive flexibility, mindfulness, and meditation play a crucial role in improving resistance to stress. The non-judgmental awareness of the present moment, or mindfulness, enables people to notice their thoughts and feelings without being overwhelmed by them (Stanley, 2014). This technique lowers stress by encouraging a balanced viewpoint on stressful situations and reducing rumination.

According to research, there are various ways in which mindfulness meditation improves resilience. First of all, it triggers the parasympathetic nervous system, which relaxes the body by counteracting the stress response (Goyal et al., 2014). Second, mindfulness training improves cognitive rigidity and enhances attention management, which enables people to adjust to changing conditions more skillfully (Geiger et al., 2016).

Additionally, practicing mindfulness improves one's emotional intelligence and capacity for self-compassion, which helps people deal with stress in a cool, collected manner. According to Stanley (2014), emotional resilience plays a crucial role in reducing the harmful effects of stress and advancing psychological health in general. Research has indicated that consistent mindfulness practice is linked to reduced levels of perceived stress and increased resilience, underscoring its effectiveness as a technique for stress management (Sharma & Rush, 2014).

Overall, mindfulness and meditation bolster resilience by promoting emotional regulation, reducing physiological stress responses, and enhancing cognitive flexibility, thereby enabling individuals to navigate and recover from stress more effectively.

10.1.2. Cognitive Behavioral Techniques

Cognitive-behavioral techniques (CBT) are instrumental in building resilience to stress by targeting maladaptive thought patterns and behaviors that contribute to stress. CBT posits that psychological distress is largely a function of dysfunctional thought processes, and by modifying these cognitions, individuals can achieve emotional and behavioral change (Macedo et al., 2014).

Identifying and addressing erroneous or irrational thoughts is a crucial aspect of cognitive behavioral therapy (CBT). Individuals can reduce the emotional impact of stressors and enhance their coping skills by swapping these out for more level-headed and sensible ideas (McCraty & Atkinson, 2012). The ability to evaluate stressful situations more precisely and less dramatically is one way that this cognitive flexibility improves resilience.

Behavioral activation, another CBT technique, encourages individuals to engage in activities that promote positive reinforcement and reduce avoidance behaviors. This engagement helps mitigate feelings of helplessness and depression, fostering a sense of accomplishment and control, which are crucial for resilience (McCraty & Atkinson, 2012).

Additionally, CBT integrates stress-reduction strategies including problem-solving and relaxation training. According to Macedo et al. (2014), these techniques give people useful tools to control and reduce stress in real-time, enhancing their general resilience. Through reorganizing unhelpful thinking patterns, encouraging adaptive behavioral adjustments, and imparting useful stress management techniques, cognitive behavioral therapy (CBT) enhances an individual's ability to control their emotions and adapt to new situations (Stanley, 2014).

10.1.3. Developing Adaptive Coping Strategies

Adaptive coping methods play a vital role in helping people develop stress resilience by helping them to effectively handle and reduce the impact of stressors. These tactics consist of finding social support, applying positive reappraisal, and overcoming problems. By recognizing stressors and creating workable solutions, problem-solving improves one's sense of efficacy and control (Southwick et al., 2014). By offering both practical and emotional help, seeking social support can improve psychological well-being and lessen feelings of loneliness (Stanley, 2014).

Positive reappraisal involves reframing a stressful situation to view it in a more positive light, fostering a sense of hope and reducing emotional distress (McCraty & Atkinson, 2012). These coping mechanisms promote psychological flexibility and adaptive functioning, which are essential components of resilience.

Adaptive coping strategies, therefore, strengthen resilience by fostering effective problem-solving, leveraging social networks, and encouraging positive cognitive reframing, thereby improving overall stress management and well-being (Macedo et al., 2014).

10.1.4. Social Support Networks

Social support networks are essential for developing stress resistance because they offer practical, emotional, and educational help. A sense of self-worth and belonging is fostered by emotional support, such as empathy, love, and trust, and this helps mitigate the harmful consequences of stress (Stanley, 2014). Informational assistance provides direction and counsel to help people deal with stressful situations in a more effective manner, improving their ability to solve problems and decreasing uncertainty (Macedo et al., 2014). Instrumental support involves tangible aid and services, such as financial assistance or help with daily tasks, alleviating the practical burdens associated with stress (Southwick et al., 2014). These various forms of support collectively contribute to a sense of security and stability, reinforcing an individual's ability to cope with and recover from stress. Research consistently demonstrates that strong social support networks are associated with lower levels of perceived stress and greater psychological resilience (McCraty & Atkinson, 2012).

10.1.5. Physical Activity and Exercise

Physical activity and exercise are essential for building resilience to stress by promoting both physiological and psychological well-being. Engaging in regular physical activity stimulates the release of endorphins, which are natural mood elevators that alleviate stress and enhance overall emotional health (Sharon-David & Tenenbaum, 2017). Exercise also reduces levels of cortisol, a stress hormone, thereby mitigating the body's stress response (Ma et al., 2017).

Additionally, exercise enhances the quality of sleep, which is essential for efficient stress reduction and recuperation (Stanković et al., 2021). In terms of psychology, exercise promotes a good self-image and increased confidence in managing stressors by giving one a sense of completion and self-efficacy. Additionally, it provides a healthy way to let go of tension and divert attention from stressful ideas, which enhances mental clarity and emotional toughness (Sharon-David & Tenenbaum, 2017).

10.2. DEVELOPING COPING SKILLS

Developing coping skills for chronic stress is essential for maintaining psychological and physiological well-being. Several evidence-based strategies can be employed to develop robust coping skills for chronic stress.

1. **Mindfulness Meditation:** A key component of mindfulness meditation is maintaining a non-judgmental concentration on the here and now. Research has demonstrated that through improving emotional regulation and lowering rumination, this exercise can lessen the symptoms of chronic stress (Goyal et al., 2014). Frequent mindfulness practice assists people in becoming more cognizant of their stress reactions and in taking a more balanced approach to stressors.
2. **Social Support:** Creating and sustaining robust social support systems is essential for managing long-term stress. Social support can mitigate the harmful consequences of stress by offering practical,

emotional, and informational help. Research indicates that those with strong social networks report feeling less stressed and having improved general health (Stanley, 2014).

3. **Cognitive-Behavioral Therapy (CBT):** Chronic stress can be effectively managed with CBT strategies including behavioral activation and cognitive restructuring. Cognitive restructuring entails recognizing and confronting unfavorable cognitive patterns, whereas behavioral activation promotes participation in enjoyable and significant activities. These methods assist people in creating more adaptable reactions to long-term stress (Macedo et al., 2014).
4. **Time Management and Organizational Skills:** Effectiveness in time management and organizing can greatly mitigate long-term stress. These abilities include prioritizing, segmenting work into digestible chunks, and effectively managing time. People can better handle the demands that lead to chronic stress by increasing productivity and decreasing feelings of overload (Twenge & Campbell, 2018).
5. **Physical Exercise:** Lower levels of chronic stress are linked to regular physical activity. Exercise raises endorphin production and lowers physiological stress markers like cortisol, which enhance mood and general well-being (Sharon-David & Tenenbaum, 2017). Another healthy way to relieve stress and enhance the quality of your sleep is to exercise.

Below are some coping skills than can be developed by applying the strategies discussed above, that are effective in navigating through stress:

1. **Thought Reframing:** CBT is among the stress and anxiety treatments with the strongest scientific backing. The idea that our thoughts affect our emotions, which in turn affect our behaviors, is the foundation of this therapeutic method. According to Macedo et al. (2014), changing the way you think about a stressor can help you better control your emotions and feel less stressed. Some tip: If you feel yourself spiraling into imagining worst-case scenarios, stop and put your mind elsewhere. Set realistic expectations for yourself. Strive for acceptance of situations outside of your control.
2. **Eliminating Stressors:** Depending on the person experiencing it as well as the situation's intensity, you may or may not endure unbearable psychological stress. Your perception and thought process regarding a stressor can also have a significant influence on your reaction (McCraty & Atkinson, 2012). While it's not always possible to avoid problems or get out of a stressful situation, you

may try to lessen the tension you're experiencing. Consider whether you can alter the stressful situation by asking for assistance, letting up of certain responsibilities, or lowering your expectations.

3. **Practicing Good Nutrition:** When confronted with a stressor, the central nervous system releases adrenaline and cortisol, which affects the digestive tract among other physiological changes. Acute stress can kill the appetite, but the release of the hormone cortisol during chronic stress can cause fat and sugar cravings. Research also suggests that high cortisol combined with high sugar consumption may prompt the deposition of fat around our internal organs-visceral fat that is associated with cardiovascular and metabolic diseases (Turner et al., 2020). A diet high in a variety of nutrients can both protect health and provide more physical energy to deal with challenges. Using substances such as alcohol to dampen the stress response does not solve the root of the problem and can have serious health effects (Yaribeygi et al., 2017).
4. **Relax Your Muscles:** Since stress causes muscle tension, being stressed out can create tension headaches, backaches, and general fatigue (Turner et al., 2020). Combat stress and these symptoms with stretches, massage, or warm baths. Or try progressive muscle relaxation, a method that has been shown to reduce anxiety and improve overall mental health (Sharon-David & Tenenbaum, 2017). To practice progressive muscle relaxation, get in a comfortable position and choose a muscle group, like your lower leg muscles (most practitioners recommend starting with the lower body and working your way up). Inhale and contract the muscles for five to 10 seconds, then exhale and release the muscles suddenly (Kaur, 2016). Relax for 10 or more seconds and then move on to the next muscle group. Another option is passive progressive muscle relaxation. This technique is similar to progressive muscle relaxation but skips the tensing step (Ma et al., 2017). Instead, simply picture each muscle group one at a time and focus on relaxing that portion of the body.
5. **Sleep Hygiene:** Daytime stress affects nighttime sleep. Making matters worse, losing shuteye can affect both cognition and mood. Try to have a consistent sleep routine that allows time to wind down. Meditation and relaxation can help with insomnia (Twenge & Campbell, 2018). Also, avoid caffeine and alcohol in the late afternoon and evening. Put down your screens, as blue light can suppress the sleepy hormone melatonin (Morales-Muñoz & Gregory, 2023). Finally, move your body during the day. A large

body of research suggests that physical activity can improve sleep, especially for middle-aged and older adults (Baranwal et al., 2023).

10.3. FINDING MEANING IN ADVERSITY

Finding meaning in adversity is a crucial aspect of psychological resilience and overall well-being. Here are some evidence-based ways individuals can derive meaning from challenging experiences:

10.3.1. Reframing and Cognitive Reappraisal

Reframing and cognitive reappraisal are crucial strategies for finding meaning in adversity by altering the interpretation of stressful events. Reframing involves changing the perspective on a situation to view it in a more positive or constructive light. This cognitive shift can transform perceived threats into challenges or opportunities for growth (Bryant & Smith, 2015). Cognitive reappraisal, a core component of emotion regulation, entails reassessing the emotional significance of an event to diminish its negative impact.

Individuals can better manage their emotions and feel in control of their situation by practicing cognitive reappraisal. Through this process, people can get insights and good meanings from negative events, which boosts resilience in addition to reducing negative emotions (Ryff, 2014). Hence, cognitive reappraisal aids in a reevaluation of one's feeling of wellbeing and purpose, assisting people in overcoming obstacles in life and moving forward.

10.3.2. Altruism and Helping Others

Devotion to the welfare of others, or altruism, is a crucial component of finding purpose in adversities. Altruistic activities provide people a sense of fulfillment and purpose, such as volunteering and helping others. By redirecting attention from one's own suffering to the beneficial effects one may have on others, this sense of purpose helps lessen the emotional toll of personal challenges (Ryff, 2014). Altruism fosters social connections and a sense of belonging, which are critical for psychological resilience. The act of helping others can lead to increased feelings of self-worth and competence, reinforcing a positive identity even in the face of adversity (Bryant & Smith, 2015). Furthermore, altruistic actions are associated with the release of endorphins and other neurochemicals that enhance mood and reduce stress, contributing to overall well-being and resilience (Macedo et al., 2014).

10.3.3. Spirituality and Religious Practices

Finding meaning in adversities is greatly facilitated by spirituality and religious practices, which offer emotional support and a framework for comprehending discomfort. Concepts of transcendence, connectivity, and purpose are common in spiritual beliefs, and they aid people in understanding and accepting their struggles (Bryant & Smith, 2015). Prayer, meditation, and group worship are examples of religious activities that help people feel more connected to their faith community and to a greater power.

These practices provide comfort and hope, enhancing emotional resilience and reducing feelings of isolation (Macedo et al., 2014). Additionally, spirituality encourages positive coping strategies, such as forgiveness, gratitude, and altruism, which further reinforce a sense of purpose and well-being. Research has shown that individuals who actively engage in spiritual and religious practices report higher levels of life satisfaction and psychological well-being, even in the face of significant adversity (Ryff, 2014).

10.3.4. Narrative Reconstruction

Individuals can reframe and integrate their experiences into a cohesive life story through narrative reconstruction, which helps them find meaning in adversity. According to Ryff (2014), this process entails organizing disorganized and upsetting memories into a more coherent and understandable framework by crafting a structured narrative that makes sense of painful or stressful moments.

Persons can develop comprehension of the reasons for and effects of their challenges by talking about what they have gone through, which helps them feel in control of their situation. According to Bryant & Smith (2015), narrative reconstruction also facilitates emotional processing by enabling people to communicate and think back on their feelings, which lessens the psychological toll that adversity takes.

Furthermore, this practice can facilitate personal growth and resilience by highlighting themes of strength, endurance, and recovery. Research indicates that individuals who engage in narrative reconstruction often report improved mental health and a greater sense of meaning and purpose in life (Macedo et al., 2014).

10.3.5. Personal Growth and Self-Discovery

Personal growth and self-discovery are pivotal in finding meaning in adversities by facilitating profound psychological transformation and resilience. When individuals face significant challenges, these experiences often prompt introspection and reevaluation of their values, priorities, and life goals (Macedo et al., 2014). This introspective process enables individuals to uncover new strengths, recognize previously unacknowledged capabilities, and develop a more nuanced understanding of themselves.

Adversities have the capacity to spur post-traumatic growth, a process in which people develop positive traits like a stronger sense of self, a better appreciation for life, and better relationships (Ryff, 2014). This transforming process gives life a deeper meaning and purpose while also lessening the detrimental effects of stress.

Self-exploration exercises, counseling, and journaling are examples of reflective practices that can help to further encourage this progress. People can gain a sense of continuity and coherence by incorporating these experiences into their self-concept, which eventually promotes resilience and well-being (Bryant & Smith, 2015).

10.4. CULTIVATING A POSITIVE MINDSET

Cultivating a positive mindset is essential for overall well-being and resilience. Here are the commonest evidence-based strategies that individuals can employ:

10.4.1. Practicing Gratitude

Practicing gratitude is a powerful way to focus on and develop appreciation for the positive things in life, which helps one develop a positive mindset. This kind of practice usually include taking regular stock of one's blessings and expressing appreciation, usually by writing in a gratitude diary or speaking aloud. Those who participated in daily thankfulness exercises reported higher levels of positive affect, more life satisfaction, and less symptoms of anxiety and depression than those who did not, according to Arslan & Wong (2024).

Gratitude practice enhances a positive mindset by shifting attention away from negative thoughts and stressors, thereby promoting a more optimistic outlook. It encourages individuals to recognize the good in their lives, which can lead to increased feelings of happiness and contentment (Cherkowski, 2018). Furthermore, gratitude fosters resilience by helping individuals find meaning and growth in difficult experiences, as it allows them to see challenges in a more positive light.

Gratitude's positive effects on social relationships bolster its psychological benefits. Gratitude can improve feelings of connectivity and social ties, both of which are essential for mental health (Cherkowski, 2018). Since people feel more appreciated and supported in their social networks, this increased social support also helps people maintain a cheerful outlook.

10.4.2. Positive Social Interactions

Positive social interactions are essential for cultivating a positive mindset by providing emotional support, fostering a sense of belonging, and enhancing overall well-being. Engaging in meaningful relationships and social activities can significantly influence one's mental health. Social support acts as a buffer against stress, mitigating its adverse effects and promoting psychological resilience (Arslan & Wong, 2024).

Engaging in positive social interactions promotes the production of oxytocin, a hormone linked to trust and bonding that lowers anxiety and increases emotions of happiness (Ryff, 2014). Furthermore, as per Macedo et al. (2014), these connections foster the exchange of affirmative experiences and reciprocal assistance, thereby strengthening an optimistic perspective on life.

Establishing and maintaining social connections also gives one the chance to practice empathy and compassion, which raises one's sense of self-worth and contentment with life. Studies reveal that those with robust social networks experience greater levels of positive affect and general well-being, which contributes to a more positive outlook (Cherkowski, 2018).

10.5. SEEKING SUPPORT IN TERMS OF STRESS

The concept of resilience, often defined as the ability to bounce back from adversity, is integral to psychological well-being and successful adaptation in the face of stress. One critical factor that enhances resilience is the ability to seek and utilize support effectively.

10.5.1. The Role of Social Support in Resilience

Social support—which includes informational, practical, and emotional support from others—is essential for reducing the negative impacts of stress and building resilience. Ozbay et al. (2007) claim that social support functions as a stress reducer by giving people access to resources that lessen the mental and practical strain of stressful circumstances. According to the study, those who feel they have a strong social support systems are less likely to experience psychological discomfort and are more likely to use adaptive coping mechanisms. Resilience requires a sense of security and belonging, both of which can be fostered by this sense of support.

The buffering hypothesis, as discussed by McCraty & Atkinson (2012), suggests that social support protects individuals from the harmful effects of stress by providing emotional comfort, practical assistance, and advice that can help reframe stressful experiences in a more manageable context. This hypothesis underscores the importance of social networks in enhancing resilience by offering tangible and intangible resources that help individuals cope more effectively with stress.

10.5.2. Emotional Support and Sturdiness

Resilience is especially aided by emotional support, which includes showing empathy, love, trust, and caring. People with higher degrees of resilience and well-being are more likely to believe they have access to emotional support. Individuals who receive emotional support are better able to manage their feelings, find meaning in their experiences, and feel less alone, all of which might worsen stress (Hubbard et al., 2018).

The development of a good self-concept and self-efficacy is facilitated by emotional support. People are more likely to think they can handle and overcome obstacles when they feel supported. This increased sense of self-efficacy is essential to resilience because it gives people the ability to confront hardship with hope and confidence.

10.5.3. Adaptation and Informational Assistance

Providing information, direction, and advice is known as informational support, and it is essential to the development of resilience. According to a study by Boszormenyi-Nagy (2014), informational support is crucial for assisting people in creating useful coping mechanisms. When people are under stress, having access to reliable information and helpful guidance can help them better comprehend their circumstances, recognize viable options, and make well-informed decisions.

Informational support is particularly valuable in situations where individuals lack the knowledge or skills to cope with specific stressors. For example, in the context of health-related stress, receiving information about treatment options, self-care strategies, and support services can significantly reduce anxiety and enhance resilience. By equipping individuals with the knowledge, they need to navigate challenges, informational support helps them feel more in control and capable of managing stress.

10.5.4. Practical Help and Essential Support

Support that is concrete and includes things like daycare, transport, and financial aid also helps people become more resilient by easing the practical responsibilities that come with stress. By easing the immediate demands on a person's resources, instrumental support enables them to concentrate on emotional and psychological coping.

Receiving financial support from family or friends during difficult times, for example, might ease the burden and stop the tension from getting worse. In a similar vein, helping people with everyday tasks can free up time and energy so they can participate in resilience-building activities like self-care. Instrumental support helps people stay stable and concentrate on long-term coping and adaptation by addressing the practical aspects of stress (Ozbay et al., 2007).

10.5.5. The Relationship Between Resilience and Types of Support

The interaction between emotional, informational, and instrumental support is just as significant as their individual functions in boosting resilience. A sturdier foundation for resilience can be provided by an all-encompassing support system that incorporates all three forms of support. When many forms of support are combined, they can have a synergistic impact where the advantages of one type of support increase the efficacy of other forms (Stanley, 2014).

For example, emotional support can enhance the impact of informational support by providing the reassurance and encouragement needed to act on advice and guidance. Similarly, instrumental support can facilitate access to emotional and informational support by reducing practical barriers. This interplay underscores the importance of a multifaceted approach to support, where individuals have access to a range of resources that address different aspects of stress and resilience.

10.5.6. Seeking Assistance as a Method to Develop Resilience

Seeking support involves more than just asking for assistance; it also entails realizing when you need it and acting to get it. Being proactive shows that a person is willing to use resources and engage in adaptive coping, which is a crucial aspect of resilience. High self-efficacy people are more likely to successfully seek out and make use of support (Glanz et al., 2008). This implies that self-efficacy-boosting therapies may also encourage behavior that seeks out help and, as a result, resilience.

Furthermore, seeking support involves building and maintaining social networks that can provide ongoing support in times of need. This requires interpersonal skills, such as communication, empathy, and reciprocity, which can be developed through social and emotional learning programs. By fostering these skills, individuals can strengthen their support networks and enhance their resilience.

10.5.7. Extending the Diversity of Your Support Network

1. **Spread Out Your Network:** It's critical to recognize that one person might not be able to meet all of your demands for social assistance. Different kinds of support can come from a variety of partnerships. A neighbor could lend a sympathetic ear when it comes

to private family matters, but a coworker might be the best person to talk to about work-related worries. Maintaining a range of solid and trustworthy relationships helps avert unfavorable encounters that could make stress worse.

2. **Take the Lead:** Waiting for others to make the first move could make you feel rejected if they don't. It is essential to actively participate in your social network. Spend time with loved ones, lend a helping hand, and stay in touch regularly. By taking the initiative, you can build stronger bonds with others and inspire reciprocation. Studies show that social support can have a substantial impact on longevity—possibly even more so than receiving it.
3. **Employ Technology:** Although in-person encounters are priceless, technology provides easy substitutes for keeping in touch with loved ones who live far away. When in-person meetings are not possible, the use of email, text messaging, and video chats can help to close the distance. Nonetheless, it's critical to strike a balance between digital and in-person contact because in-person communication is frequently more advantageous.
4. **Participate in Common Interests:** Developing ties with people who have similar interests to your own can be a successful strategy. Playing sports, hiking, singing, making, or being involved in local politics are just a few examples of activities that you can get others who share your interests with. As deep friendships grow over time by shared experiences, patience is essential.
5. **Seek Peer Help:** In certain stressful situations, including managing a chronic disease or providing care for someone else, you may need help from people outside of your current network. During these moments, joining a support group can offer the specialized help and understanding that's needed.
6. **Enhance Social Skills:** If social interactions feel awkward or intimidating, initiating conversations with simple questions about the other person can ease the process. Engaging in shared activities, such as group hobbies or classes, can be less daunting than unstructured socializing. For those experiencing significant social anxiety, consulting a therapist skilled in social anxiety and social skills training can be beneficial (Glanz et al., 2008).

Seeking support is a crucial component of resilience in the face of stress. Emotional, informational, and instrumental support each plays distinct roles in enhancing resilience by providing resources that alleviate the emotional, cognitive, and practical burdens of stress. The interplay of different types of support creates a comprehensive support system that can more effectively buffer against stress and promote resilience. By recognizing the importance of support and taking proactive steps to seek and utilize it, individuals can build and maintain the resilience needed to navigate the challenges of life.

10.6. PRACTICING SELF-CARE

Self-care is a complex concept that includes a range of actions and methods intended to enhance general well-being. It is becoming more widely acknowledged as being essential to building resilience against stress.

- **Integrating Resilience and Self-Care:** Self-care techniques include everything from psychological techniques like mindfulness and reaching out for social support to physical activities like regular exercise and rest. These routines are essential for preserving homeostasis and lessening the damaging effects of stress. We'll go into detail in this part on maintaining self-care routines as a means of managing stress.

10.6.1. Importance of Physical Self-Care

Exercise, a healthy diet, and adequate sleep are examples of physical self-care. These practices are fundamental to stress management and resilience building. According to a study by Gerber et al. (2014), exercise helps prevent health problems linked to stress. Regular physical exercise, according to scientists, greatly lessens the negative consequences of stress on mental health, including symptoms of despair and anxiety. Exercise improves sleep quality, which is essential for both emotional control and cognitive performance and encourages the release of endorphins, which are naturally occurring mood enhancers.

Moreover, healthy eating is a necessary component of physical self-care. Stress management is greatly aided by nutritional self-care techniques like eating a balanced diet full of vital nutrients. O'Neil et al. (2014) found a significant relationship between food quality and outcomes related to mental health. According to the scientists, eating a diet rich in nutrients can improve psychological resilience overall and lower the chance of mental health illnesses arising. Eating a healthy diet gives the brain the energy and nutrients it needs to function at its best, which makes it easier for people to handle stress.

10.6.2. Mental Self-Care and Mindfulness

Building resilience against stress requires psychological self-care, which includes techniques like mindfulness, meditation, and cognitive behavioral tactics. It has been demonstrated that practicing mindfulness, which is the nonjudgmental awareness of the present moment, greatly lowers stress and improves emotional control. According to a study by Glanz et al. (2008), mindfulness-based stress reduction (MBSR) programs effectively reduce anxiety and stress symptoms, which in turn increases resilience. The study's findings revealed that individuals who practiced mindfulness had reduced levels of the stress hormone cortisol and reported feeling better emotionally.

Furthermore, psychological self-care includes cognitive strategies such as positive reframing and problem-solving. These strategies enable individuals to view stressors from a different perspective and develop effective coping mechanisms. The ability to reframe negative thoughts and focus on solutions rather than problems is a hallmark of resilient individuals. This cognitive flexibility, fostered through self-care practices, allows individuals to adapt to changing circumstances and recover from setbacks more efficiently.

10.6.3. Social Support as a Component of Self-Care

Building resilience also requires social self-care, which is establishing and sustaining relationships with helpful people. In order to effectively manage stress, social support networks offer a sense of belonging, practical help, and emotional support. Ozbay et al. (2007) found that reduced stress levels and better mental health outcomes are linked to robust social support. The authors contend that social contacts, which offer people real resources, guidance, and emotional support, can mitigate the negative effects of stress.

Moreover, social support facilitates the sharing of experiences and coping strategies, which can enhance resilience. Individuals who engage in social self-care by nurturing their relationships and seeking support when needed are better equipped to handle stressors. These social connections provide a sense of security and validation, which are crucial for maintaining mental well-being.

10.6.4. An Integrated Method for Self-Care

To build resilience against stress, an integrative approach to self-care that incorporates social, psychological, and physical strategies works well. This all-encompassing approach encourages a thorough approach to stress management while acknowledging the connections between various facets of well-being. For example, getting regular exercise, meditating, and reaching out to others can all work together to improve resilience. Each element of self-care strengthens the others, resulting in a strong foundation for managing stress.

10.6.5. Consequences for Clinical Practice

The knowledge gained from these academic publications emphasizes how crucial it is to support self-care behaviors in a variety of contexts, such as the workplace, classrooms, and hospital settings. Employers and educators have a critical role to play in building resilience by promoting self-care among staff members and pupils. This can be accomplished by implementing programs like mindfulness training, wellness centers, and welcoming places that put mental health first.

In their practice, healthcare providers should also stress the value of self-care. Healthcare professionals can assist people in developing resilience and enhancing their general well-being by teaching patients about the advantages of self-care and offering resources to support these activities. Including self-care in treatment regimens can improve therapies' effectiveness and support mental health over the long term.

10.7. EMBRACING CHANGE AND GROWTH

Change is an inherent aspect of life, and the ability to adapt to change is crucial for resilience. In their study, Southwick et al. (2014) highlight that resilience is not a static trait but a dynamic process that involves positive adaptation within the context of significant adversity. They argue that individuals who view change as an opportunity for growth rather than a threat are more likely to develop resilience. This perspective allows them to reframe stressful situations, perceive them as challenges to be overcome, and thus maintain a positive outlook. Southwick et al. emphasize that cognitive flexibility, the mental ability to switch between thinking about two different concepts or thinking about multiple concepts simultaneously, is crucial in this process. It enables individuals to adapt to new situations and challenges, facilitating resilience.

Furthermore, accepting change means accepting the uncertainty that it entails. According to Forbes & Fikretoglu's (2018) research, those who embrace change and uncertainty as a natural part of life are better able to employ happy feelings as a coping mechanism when faced with stress. According to their broaden-and-build hypothesis of positive emotions, experiencing happy feelings can help someone become more aware of their surroundings and inspire them to think and behave in new, creative, and experimental ways. This expansion creates long-lasting personal assets over time, such as heightened resilience. Thus, accepting change promotes long-term resilience in addition to helping people deal with current stress.

10.7.1. Personal Development as a Resilience Mechanism

Personal development, which is frequently brought about by encountering and overcoming hardship, is another essential element of resilience. Progress entails learning new abilities, obtaining fresh perspectives, and reaching a more profound comprehension of oneself and the universe. People who go through a lot of stress or trauma frequently talk about how their lives have improved in positive ways, such their relationships, their respect for life, and their sense of personal strength. These modifications, which are a sign of personal development, help people become more resilient by arming them with the mental tools they need to handle challenges in the future.

According to Tedeschi & Calhoun's (2004) post-traumatic growth model, cognitive processing—actively making sense of and deriving meaning from stressful experiences—is essential to human development. People who participate in this kind of cognitive processing are more likely to grow personally and, as a result, become resilient. Identifying lessons learned from experiences and incorporating them into one's life story are common steps in this approach. By doing this, people learn from their experiences and improve their ability to face obstacles in the future.

The development of resilience through personal growth is supported by the concept of self-efficacy. Self-efficacy, the belief in one's ability to influence events and outcomes in one's life, is enhanced through the successful navigation of stress and adversity. When individuals face challenges and emerge stronger, their confidence in their ability to manage future stressors increases. This enhanced self-efficacy contributes to resilience by fostering a proactive approach to problem-solving and stress management.

10.7.2. Integrating Growth and Change for Resilience

A complete strategy for developing resilience is offered by combining the ideas of welcoming change and encouraging personal development. These two components work together to produce a synergistic effect that improves a person's ability to handle stress. According to Southwick et al. (2014), while good emotional experiences expand one's thought-action repertoire and help one build personal resources, cognitive flexibility enables one to adjust to changing situations.

Tedeschi & Calhoun (2004) assert that positive emotions not only help individuals cope with stress in the short term but also contribute to long-term resilience by building enduring personal resources. These resources, such as increased social support, enhanced coping skills, and greater psychological well-being, provide a foundation for personal growth. As individuals grow from their experiences, they become more resilient, creating a positive feedback loop that reinforces their ability to embrace change and continue growing.

The significance of deriving meaning from adversity is emphasized by Tedeschi & Calhoun's (2004) research on post-traumatic growth. People develop a better feeling of resilience when they can make sense of their experiences and recognize their improvement. This procedure is taking stock of one's experiences, comprehending the lessons discovered, and incorporating those lessons into one's personal story. By doing this, people learn from their experiences and improve their ability to face obstacles in the future.

10.7.3. Practical Implications

It is possible to improve resilience in a variety of situations, such as the job, school, and mental health treatments, by realizing the value of welcoming change and encouraging personal development. Resilience training, which prioritizes self-efficacy, positive emotion control, and cognitive flexibility, can be incorporated into educational programs. These kinds of programs can assist students in gaining the abilities necessary to adjust to shifting conditions and see obstacles as chances for personal development.

In the workplace, organizations can foster a culture of resilience by encouraging employees to embrace change and pursue personal and professional growth. This can be achieved through training programs, mentorship, and creating an environment that supports risk-taking and learning from failure. By promoting a growth mindset, organizations can help employees build resilience and enhance overall organizational performance.

Strategies that build resilience by accepting change and encouraging personal development can also be beneficially included into mental health interventions. Individuals can acquire the psychological tools necessary to manage stress and hardship by focusing on cognitive flexibility, positive emotion control, and self-efficacy in cognitive-behavioral therapy. Furthermore, interventions that support people in thinking back on their past, finding purpose in hardship, and identifying areas of personal development can strengthen resilience and lead to better mental health outcomes.

BIBLIOGRAPHY

1. Amarya, S., Singh, K., & Sabharwal, M. (2018). Ageing process and physiological changes. In G. D'Onofrio, A. Greco, & D. Sancarlo (Eds.), *Gerontology*. InTech. https://doi.org/10.5772/intechopen.76249.
2. American Psychological Association. (2020). *Stress in America: A National Mental Health Crisis*. https://www.apa.org/news/press/releases/stress/2020/report-october (accessed on 21 August 2024).
3. Arslan, G., & Wong, P. T. (2024). Embracing life's challenges: Developing a tool for assessing resilient mindset in second wave positive psychology. *Journal of Happiness and Health, 4*(1), Article 1. https://doi.org/10.47602/johah.v4i1.53.
4. Balters, S., Geeseman, J. W., Tveten, A. K., Hildre, H. P., Ju, W., & Steinert, M. (2020). Mayday, mayday, mayday: Using salivary cortisol to detect distress (and eustress!) in critical incident training. *International Journal of Industrial Ergonomics, 78*, 102975. https://doi.org/10.1016/j.ergon.2020.102975.
5. Baranwal, N., Yu, P. K., & Siegel, N. S. (2023). Sleep physiology, pathophysiology, and sleep hygiene. *Progress in Cardiovascular Diseases, 77*, 59–69. https://doi.org/10.1016/j.pcad.2023.02.005.
6. Barrero-Castillero, A., Morton, S. U., Nelson, C. A., & Smith, V. C. (2019). Psychosocial stress and adversity: Effects from the perinatal period to adulthood. *NeoReviews, 20*(12), e686–e696. https://doi.org/10.1542/neo.20-12-e686.
7. Barsotti, N., Chiera, M., Lanaro, D., & Fioranelli, M. (2020). Impact of stress, immunity, and signals from endocrine and nervous system on fascia. *Frontiers in Bioscience-Elite, 13*(1), Article 1. https://doi.org/10.2741/870.

8. Baum, A. (1984). *Handbook of Psychology and Health: Social Psychological Aspects of Health*. Erlbaum.
9. Berndt, J., & Ortelli, T. A. (2023). Creating a healthy work environment. *123*(3), 59–60. https://doi.org/10.1097/01.NAJ.0000921820.57044.2b.
10. Birch-Machin, M. A., & Bowman, A. (2016). Oxidative stress and ageing. *British Journal of Dermatology, 175*, 26–29. https://doi.org/10.1111/bjd.14906.
11. Bonnar, D., Bartel, K., Kakoschke, N., & Lang, C. (2018). Sleep interventions designed to improve athletic performance and recovery: A systematic review of current approaches. *Sports Medicine, 48*(3), 683–703. https://doi.org/10.1007/s40279-017-0832-x.
12. Brunges, M., & Foley-Brinza, C. (2014). Projects for increasing job satisfaction and creating a healthy work environment. *AORN Journal, 100*(6), 670–681. https://doi.org/10.1016/j.aorn.2014.01.029.
13. Brunt, D. (2023). Stress and digestion: 10 things you should know. *Otepoti Integrative*. https://www.otepotiintegrativehealth.co.nz/post/stress-and-digestion-10-things-you-should-know (accessed on 21 August 2024).
14. Bryant, F. B., & Smith, J. L. (2015). Appreciating life in the midst of adversity: Savoring in relation to mindfulness, reappraisal, and meaning. *Psychological Inquiry, 26*(4), 315–321. https://doi.org/10.1080/1047840X.2015.1075351.
15. Buysse, D. J. (2014). Sleep health: Can we define it? Does it matter? *Sleep, 37*(1), 9–17. https://doi.org/10.5665/sleep.3298.
16. Campbell, J., & Ehlert, U. (2012). Acute psychosocial stress: Does the emotional stress response correspond with physiological responses? *Psych Neuroendocrinology, 37*(8), 1111–1134. https://doi.org/10.1016/j.psyneuen.2011.12.010.
17. Chang, A. M., Aeschbach, D., Duffy, J. F., & Czeisler, C. A. (2015). Evening use of light-emitting eReaders negatively affects sleep, circadian timing, and next-morning alertness. *Proceedings of the National Academy of Sciences, 112*(4), 1232–1237. https://doi.org/10.1073/pnas.1418490112.
18. Chawla, L., Keena, K., Pevec, I., & Stanley, E. (2014). Green schoolyards as havens from stress and resources for resilience in childhood and adolescence. *Health & Place, 28*, 1–13. https://doi.org/10.1016/j.healthplace.2014.03.001.
19. Cherkowski, S. (2018). Positive teacher leadership: Building mindsets and capacities to grow wellbeing. *International Journal of Teacher Leadership, 9*(1), 63–78.
20. Chu, B., Marwaha, K., Sanvictores, T., Awosika, A. O., & Ayers, D.

(2024). Physiology, stress reaction. In *StatPearls*. StatPearls Publishing. http://www.ncbi.nlm.nih.gov/books/NBK541120/ (accessed on 21 August 2024).

21. Cohen, S., Gianaros, P. J., & Manuck, S. B. (2016). A stage model of stress and disease. *Perspectives on Psychological Science*. https://doi.org/10.1177/1745691616646305.
22. Compas, B. E., Jaser, S. S., Bettis, A. H., Watson, K. H., Gruhn, M. A., Dunbar, J. P., Williams, E., & Thigpen, J. C. (2017). Coping, emotion regulation, and psychopathology in childhood and adolescence: A meta-analysis and narrative review. *Psychological Bulletin, 143*(9), 939–991. https://doi.org/10.1037/bul0000110.
23. Cooper, E. (2023). Why you should bake to relieve stress. *Emma Cooper*. https://emmacooper.org/why-you-should-bake-to-relieve-stress/ (accessed on 21 August 2024).
24. Cordell, R. C. (2024). Is stress impacting your sleep? Check out these tips for better rest when you're feeling overwhelmed. *Sleep Advisor*. Retrieved from: https://www.sleepadvisor.org/stress-and-sleep/ (accessed on 21 August 2024).
25. Cummings, J., Lee, G., Ritter, A., Sabbagh, M., & Zhong, K. (2020). Alzheimer's disease drug development pipeline: 2020. *Alzheimer's & Dementia: Translational Research & Clinical Interventions, 6*(1), e12050. https://doi.org/10.1002/trc2.12050.
26. de Magalhães, J. P., & Passos, J. F. (2018). Stress, cell senescence and organismal ageing. *Mechanisms of Ageing and Development, 170*, 2–9. https://doi.org/10.1016/j.mad.2017.07.001.
27. De Witte, N. A. J., Joris, S., Van Assche, E., & Van Daele, T. (2021). Technological and digital interventions for mental health and wellbeing: An overview of systematic reviews. *Frontiers in Digital Health, 3*, 754337. https://doi.org/10.3389/fdgth.2021.754337.
28. Dhabhar, F. S. (2014). Effects of stress on immune function: The good, the bad, and the beautiful. *Immunologic Research, 58*(2), 193–210. https://doi.org/10.1007/s12026-014-8517-0.
29. Dorsey, M. D. D. (2022). Does stress really affect your immunity, or is that just hippy nonsense? *CHOQ®*. Retrieved from: https://choq.com/does-stress-really-affect-your-immunity-or-is-that-just-hippy-nonsense/ (accessed on 21 August 2024).
30. Dray, J. (2021). Child and adolescent mental health and resilience-focused interventions: A conceptual analysis to inform future research. *International Journal of Environmental Research and Public Health, 18*(14), 7315. https://doi.org/10.3390/ijerph18147315.

31. Durlak, J. A., Weissberg, R. P., Dymnicki, A. B., Taylor, R. D., & Schellinger, K. B. (2011). The impact of enhancing students' social and emotional learning: A meta-analysis of school-based universal interventions. *Child Development, 82*(1), 405–432. https://doi.org/10.1111/j.1467-8624.2010.01564.x.

32. Eby, K. (2023). How to improve work-life balance: 17 tips for success. *Smartsheet.* Retrieved from: https://www.smartsheet.com/content/how-to-improve-work-life-balance (accessed on 21 August 2024).

33. Economides, M., Martman, J., Bell, M. J., & Sanderson, B. (2018). Improvements in stress, affect, and irritability following brief use of a mindfulness-based smartphone app: A randomized controlled trial. *Mindfulness, 9*(5), 1584–1593. https://doi.org/10.1007/s12671-018-0905-4.

34. Fletcher, M. (2021). 3 types of stress and how the body responds. *Anticancer Lifestyle Program.* https://anticancerlifestyle.org/3-types-of-stress-and-how-the-body-responds/ (accessed on 21 August 2024).

35. Forbes, S., & Fikretoglu, D. (2018). Building resilience: The conceptual basis and research evidence for resilience training programs. *Review of General Psychology, 22*(4), 452–468. https://doi.org/10.1037/gpr0000152.

36. Formen Health. (2024). How stress affects your relationships and health. Retrieved from: https://www.formen.health/blogs/health-wellness/stress-ruining-relationships (accessed on 21 August 2024).

37. Foy, T., Dwyer, R. J., Nafarrete, R., Hammoud, M. S. S., & Rockett, P. (2019). Managing job performance, social support and work-life conflict to reduce workplace stress. *International Journal of Productivity and Performance Management, 68*(6), 1018–1041. https://doi.org/10.1108/IJPPM-03-2017-0061.

38. Galluch, P., Grover, V., & Thatcher, J. (2015). Interrupting the workplace: Examining stressors in an information technology context. *Journal of the Association for Information Systems, 16*(1), 1–47. https://doi.org/10.17705/1jais.00387.

39. Geiger, P. J., Boggero, I. A., Brake, C. A., Caldera, C. A., Combs, H. L., Peters, J. R., & Bacr, R. A. (2016). Mindfulness-based interventions for older adults: A review of the effects on physical and emotional well-being. *Mindfulness, 7*(2), 296–307. https://doi.org/10.1007/s12671-015-0444-1.

40. Geiker, N. R. W., Astrup, A., Hjorth, M. F., Sjödin, A., Pijls, L., & Markus, C. R. (2018). Does stress influence sleep patterns, food intake, weight gain, abdominal obesity and weight loss interventions and vice versa? *Obesity Reviews, 19*(1), 81–97. https://doi.org/10.1111/obr.12603.

41. Gemma, C., Vila, J., Bachstetter, A., & Bickford, P. C. (2007). Oxidative stress and the aging brain: From theory to prevention. In D. R. Riddle (Ed.), *Brain Aging: Models, Methods, and Mechanisms*. CRC Press/Taylor & Francis. http://www.ncbi.nlm.nih.gov/books/NBK3869/ (accessed on 21 August 2024).

42. Glanz, K., Rimer, B. K., & Viswanath, K. (2008). *Health Behavior and Health Education: Theory, Research, and Practice* (4th ed.). Jossey-Bass.

43. Glazer, S., & Liu, C. (2017). Work, stress, coping, and stress management. In *Oxford Research Encyclopedia of Psychology*. https://oxfordre.com/psychology/psychology/view/10.1093/acrefore/9780190236557.001.0001/acrefore-9780190236557-e-30.

44. Goltzman, J. (2023). 25 tips and techniques to try for couple's therapy. *Healthline*. https://www.healthline.com/health/mental-health/therapy-for-couples (accessed on 21 August 2024).

45. Goyal, M., Singh, S., Sibinga, E. M. S., Gould, N. F., Rowland-Seymour, A., Sharma, R., Berger, Z., Sleicher, D., Maron, D. D., Shihab, H. M., Ranasinghe, P. D., Linn, S., Saha, S., Bass, E. B., & Haythornthwaite, J. A. (2014). Meditation programs for psychological stress and well-being: A systematic review and meta-analysis. *JAMA Internal Medicine, 174*(3), 357–368. https://doi.org/10.1001/jamainternmed.2013.13018.

46. Greensboro Chiropractic. (2020). *Body signals: Stress*. Retrieved from: https://www.greensborochiropractor.net/body-signals-stress/ (accessed on 21 August 2024).

47. Harvard Health. (2020). Exercising to relax. Retrieved from: https://www.health.harvard.edu/staying-healthy/exercising-to-relax (accessed on 21 August 2024).

48. Hasan, A. A., & Tumah, H. (2019). The correlation between occupational stress, coping strategies, and the levels of psychological distress among nurses working in mental health hospitals in Jordan. *Perspectives in Psychiatric Care, 55*(2), 153–160. https://doi.org/10.1111/ppc.12292.

49. Hill, D., Conner, M., Clancy, F., Moss, R., Wilding, S., Bristow, M., & O'Connor, D. B. (2022). Stress and eating behaviors in healthy adults: A systematic review and meta-analysis. *Health Psychology Review, 16*(2), 280–304. https://doi.org/10.1080/17437199.2021.1923406.

50. I Got Mind Inc. (2023). Course on understanding stress presented by I Got Mind. Retrieved from: https://igotmind.ca/course/understanding-stress-for-business/ (accessed on 21 August 2024).

51. Kalmbach, D. A., Anderson, J. R., & Drake, C. L. (2018). The impact of stress on sleep: Pathogenic sleep reactivity as a vulnerability to insomnia

and circadian disorders. *Journal of Sleep Research, 27*(6), e12710. https://doi.org/10.1111/jsr.12710.

52. Kaur, J. (2016). Efficacy of progressive relaxation training on anxiety of adolescents. *International Journal of Physical Education, Sports and Health.*

53. Kevin-Cleary. (2023). What is progressive muscle relaxation? *Health Products for You.* https://www.healthproductsforyou.com/ar-what-is-progressive-muscle-relaxation.html (accessed on 21 August 2024).

54. Khoury, B., Sharma, M., Rush, S. E., & Fournier, C. (2015). Mindfulness-based stress reduction for healthy individuals: A meta-analysis. *Journal of Psychosomatic Research, 78*(6), 519–528. https://doi.org/10.1016/j.jpsychores.2015.03.009.

55. Kiddipedia. (2024). Is my child stressed? *Kiddipedia.* Retrieved from: https://kiddipedia.com.au/is-my-child-stressed/ (accessed on 21 August 2024).

56. Kiecolt-Glaser, J. K., & Wilson, S. J. (2017). Lovesick: How couples' relationships influence health. *Annual Review of Clinical Psychology, 13*(1), 421–443. https://doi.org/10.1146/annurev-clinpsy-032816-045111.

57. Kivimäki, M., & Steptoe, A. (2018). Effects of stress on the development and progression of cardiovascular disease. *Nature Reviews Cardiology, 15*(4), 215–229. https://doi.org/10.1038/nrcardio.2017.189.

58. KlearMinds. (2024). *CBT for Stress Management.* https://www.klearminds.com/therapies/cbt-cognitive-behavioural-therapy/stress/ (accessed on 21 August 2024).

59. Lee, A. R., Son, S. M., & Kim, K. K. (2016). Information and communication technology overload and social networking service fatigue: A stress perspective. *Computers in Human Behavior, 55*, 51–61. https://doi.org/10.1016/j.chb.2015.08.011.

60. Liu, Y. Z., Wang, Y. X., & Jiang, C. L. (2017). Inflammation: The common pathway of stress-related diseases. *Frontiers in Human Neuroscience, 11.* https://doi.org/10.3389/fnhum.2017.00316.

61. Lockett, E. L. (2023). Your guide to understanding and managing workplace stress. *Healthline.* Retrieved from: https://www.healthline.com/health/stress/workplace-stress (accessed on 21 August 2024).

62. Loi, M., Mossink, J. C. L., Meerhoff, G. F., Den Blaauwen, J. L., Lucassen, P. J., & Joëls, M. (2017). Effects of early-life stress on cognitive function and hippocampal structure in female rodents. *Neuroscience, 342*, 101–119. https://doi.org/10.1016/j.neuroscience.2015.08.024.

63. Longevity. (2021). The link between stress and aging: How reversing its

effects is possible. Retrieved from: https://vocal.media/longevity/the-link-between-stress-and-aging-how-reversing-its-effects-is-possible (accessed on 21 August 2024).

64. Lundqvist, D., Fogelberg Eriksson, A., & Ekberg, K. (2018). Managers' social support: Facilitators and hindrances for seeking support at work. *Work, 59*(3), 351–365. https://doi.org/10.3233/WOR-182690.

65. Ma, X., Yue, Z. Q., Gong, Z. Q., Zhang, H., Duan, N. Y., Shi, Y. T., Wei, G. X., & Li, Y. F. (2017). The effect of diaphragmatic breathing on attention, negative affect and stress in healthy adults. *Frontiers in Psychology, 8*, 874. https://doi.org/10.3389/fpsyg.2017.00874.

66. Macedo, T., Wilheim, L., Gonçalves, R., Coutinho, E. S. F., Vilete, L., Figueira, I., & Ventura, P. (2014). Building resilience for future adversity: A systematic review of interventions in non-clinical samples of adults. *BMC Psychiatry, 14*(1), 227. https://doi.org/10.1186/s12888-014-0227-6.

67. Maier, C., Laumer, S., Weinert, C., & Weitzel, T. (2015). The effects of technostress and switching stress on discontinued use of social networking services: A study of Facebook use. *Information Systems Journal, 25*(3), 275–308. https://doi.org/10.1111/isj.12068.

68. Masarik, A. S., & Conger, R. D. (2017). Stress and child development: A review of the family stress model. *Current Opinion in Psychology, 13*, 85–90. https://doi.org/10.1016/j.copsyc.2016.05.008.

69. Masson, P. (2023). Employee burnout! The blindspot for business leaders. *LinkedIn*. https://www.linkedin.com/pulse/employee-burnout-blindspot-business-leaders-philippe-masson/ (accessed on 21 August 2024).

70. McCraty, R., & Atkinson, M. (2012). Resilience training program reduces physiological and psychological stress in police officers. *Global Advances in Health and Medicine*. https://doi.org/10.7453/gahmj.2012.1.5.013.

71. McEwen, B. S. (2017). Neurobiological and systemic effects of chronic stress. *Chronic Stress, 1*, 2470547017692328. https://doi.org/10.1177/2470547017692328.

72. McLaughlin, K. A., Sheridan, M. A., Tibu, F., Fox, N. A., Zeanah, C. H., & Nelson, C. A. (2015). Causal effects of the early caregiving environment on development of stress response systems in children. *Proceedings of the National Academy of Sciences, 112*(18), 5637–5642. https://doi.org/10.1073/pnas.1423363112.

73. Monin, J. K., Levy, B. R., & Kane, H. S. (2017). To love is to suffer: Older adults' daily emotional contagion to perceived spousal suffering. *The Journals of Gerontology: Series B, 72*(3), 383–387. https://doi.org/10.1093/geronb/gbv070.

74. Morales-Muñoz, I., & Gregory, A. M. (2023). Sleep and mental health problems in children and adolescents. *Sleep Medicine Clinics, 18*(2), 245–254. https://doi.org/10.1016/j.jsmc.2023.01.006.

75. Morera, L. P., Marchiori, G. N., Medrano, L. A., & Defagó, M. D. (2019). Stress, dietary patterns and cardiovascular disease: A mini-review. *Frontiers in Neuroscience, 13*. https://doi.org/10.3389/fnins.2019.01226.

76. Nijjar, P. S., Connett, J. E., Lindquist, R., Brown, R., Burt, M., Pergolski, A., Wolfe, A., Balaji, P., Chandiramani, N., Yu, X., Kreitzer, M. J., & Everson-Rose, S. A. (2019). Randomized trial of mindfulness-based stress reduction in cardiac patients eligible for cardiac rehabilitation. *Scientific Reports, 9*(1). https://doi.org/10.1038/s41598-019-54932-2.

77. Overall, N. C., & McNulty, J. K. (2017). What type of communication during conflict is beneficial for intimate relationships? *Current Opinion in Psychology, 13*, 1–5. https://doi.org/10.1016/j.copsyc.2016.03.002.

78. Ovington, T. (2023). What is technostress? (Types + how to reduce it). *WalkMe Blog*. https://www.walkme.com/blog/technostress/ (accessed on 21 August 2024).

79. Perrier, M. F., Gurgel-Juarez, N., Flowers, H. L., McCormick, A., & Short, S. J. (2020). Mindfulness-based interventions for children and adolescents across all settings: A scoping review protocol. *Systematic Reviews, 9*(1), 286. https://doi.org/10.1186/s13643-020-01548-7.

80. Randall, A. K., & Bodenmann, G. (2017). Stress and its associations with relationship satisfaction. *Current Opinion in Psychology, 13*, 96–106. https://doi.org/10.1016/j.copsyc.2016.05.010.

81. Robby. (2021). How stress affects children's health and what you can do to protect them. *Dr. Lam Coaching–World Renowned Authority on Adrenal Fatigue Recovery*. https://www.drlamcoaching.com/blog/stress-affects-childrens-health/ (accessed on 21 August 2024).

82. Rossiello, F., Jurk, D., Passos, J. F., & Di Fagagna, F. D. (2022). Telomere dysfunction in ageing and age-related diseases. *Nature Cell Biology, 24*(2), 135–147. https://doi.org/10.1038/s41556-022-00842-x.

83. Russell, L. (2024). Stress in children: Powerful action steps for parents [+ free PDF guide]. *They Are the Future*. https://www.theyarethefuture.co.uk/how-stressed-is-your-child/ (accessed on 21 August 2024).

84. Ryff, C. D. (2014). Self-realization and meaning making in the face of adversity: A eudaimonic approach to human resilience. *Journal of Psychology in Africa, 24*(1), 1–12. https://doi.org/10.1080/14330237.2014.904098.

85. Sadeghi, P., Pezeshki, P. S., & Rezaei, N. (2023). Coronavirus disease 2019

(COVID-19) in pediatric patients with autoimmune disorders. *European Journal of Pediatrics, 182*(7), 2967–2988. https://doi.org/10.1007/s00431-023-04958-6.

86. Schoenberg, P. L. A., & David, A. S. (2014). Biofeedback for psychiatric disorders: A systematic review. *Applied Psychophysiology and Biofeedback, 39*(2), 109–135. https://doi.org/10.1007/s10484-014-9246-9.
87. Scott, A. J., Webb, T. L., Martyn-St James, M., Rowse, G., & Weich, S. (2021). Improving sleep quality leads to better mental health: A meta-analysis of randomized controlled trials. *Sleep Medicine Reviews, 60,* 101556. https://doi.org/10.1016/j.smrv.2021.101556.
88. Shalev, A. Y. (2009). Posttraumatic stress disorder and stress-related disorders. *The Psychiatric Clinics of North America, 32*(3), 687–704. https://doi.org/10.1016/j.psc.2009.06.001.
89. Shamdani, S. (2022). Constructive & collaborative problem solving: Tools for couples going through a separation process. *LinkedIn*. https://www.linkedin.com/pulse/constructive-collaborative-problem-solving-tools-sara-shamdani-phd/ (accessed on 21 August 2024).
90. Sharma, M., & Rush, S. E. (2014). Mindfulness-based stress reduction as a stress management intervention for healthy individuals: A systematic review. *Journal of Evidence-Based Complementary & Alternative Medicine, 19*(4), 271–286. https://doi.org/10.1177/2156587214543143.
91. Sharon-David, H., & Tenenbaum, G. (2017). The effectiveness of exercise interventions on coping with stress: Research synthesis. *Studies in Sport Humanities, 22*, 19–29. https://doi.org/10.5604/01.3001.0012.6520.
92. Simpson, J. A., & Rholes, W. S. (2017). Adult attachment, stress, and romantic relationships. *Current Opinion in Psychology, 13*, 19–24. https://doi.org/10.1016/j.copsyc.2016.04.006.
93. Solomon, D. H., Knobloch, L. K., Theiss, J. A., & McLaren, R. M. (2016). Relational turbulence theory: Explaining variation in subjective experiences and communication within romantic relationships. *Human Communication Research, 42*(4), 507–532. https://doi.org/10.1111/hcre.12091.
94. Southwick, S. M., Bonanno, G. A., Masten, A. S., Panter-Brick, C., & Yehuda, R. (2014). Resilience definitions, theory, and challenges: Interdisciplinary perspectives. *European Journal of Psychotraumatology, 5*, 25338. https://doi.org/10.3402/ejpt.v5.25338.
95. Stanković, M., Nešić, M., Čičević, S., & Shi, Z. (2021). Association of smartphone use with depression, anxiety, stress, sleep quality, and internet

addiction: Empirical evidence from a smartphone application. *Personality and Individual Differences, 168*, 110342. https://doi.org/10.1016/j.paid.2020.110342.

96. Stanley, E. A. (2014). Mindfulness-based mind fitness training: An approach for enhancing performance and building resilience in high-stress contexts. In *The Wiley Blackwell Handbook of Mindfulness* (pp. 964–985). John Wiley & Sons, Ltd. https://doi.org/10.1002/9781118294895.ch50.
97. Suni, E., & Dimitriu, A. (2024). What causes insomnia? *Sleep Foundation*. Retrieved from: https://www.sleepfoundation.org/insomnia/what-causes-insomnia (accessed on 21 August 2024).
98. Syvertsen, T., & Enli, G. (2020). Digital detox: Media resistance and the promise of authenticity. *Convergence: The International Journal of Research into New Media Technologies, 26*(5–6), 1269–1283. https://doi.org/10.1177/1354856519847325.
99. Taylor, T. H. (2018). Stress and social media. *LinkedIn*. https://www.linkedin.com/pulse/stress-social-media-tish-holub-taylor/ (accessed on 21 August 2024).
100. Tran, C. T. H., Tran, H. T. M., Nguyen, H. T. N., Mach, D. N., Phan, H. S. P., & Mujtaba, B. G. (2020). Stress management in the modern workplace and the role of human resource professionals. *Business Ethics and Leadership, 4*(2), 26–40. https://doi.org/10.21272/bel.4(2).26-40.2020.
101. Turner, A. I., Smyth, N., Hall, S. J., Torres, S. J., Hussein, M., Jayasinghe, S. U., Ball, K., & Clow, A. J. (2020). Psychological stress reactivity and future health and disease outcomes: A systematic review of prospective evidence. *Psych Neuroendocrinology, 114*, 104599. https://doi.org/10.1016/j.psyneuen.2020.104599.
102. Twenge, J. M., & Campbell, W. K. (2018). Associations between screen time and lower psychological well-being among children and adolescents: Evidence from a population-based study. *Preventive Medicine Reports, 12*, 271–283. https://doi.org/10.1016/j.pmedr.2018.10.003.
103. Williams, W. M., Barnett, S. M., & Wethington, E. (2015). Balancing professional and personal life. In *Success Strategies from Women in STEM* (pp. 391–411). Elsevier. https://doi.org/10.1016/B978-0-12-397181-4.00012-8.
104. Wise, V. (2022). Treating a sleep disorder with therapy yields lasting results. *HealthMatch*. https://healthmatch.io/blog/science-shows-treating-a-sleep-disorder-with-therapy-yields-lasting-results (accessed on 21 August 2024).
105. Wunsch, K., Meier, M., Ueberholz, L., Strahler, J., & Kasten, N. (2019). Acute psychosocial stress and working memory performance: The

potential of physical activity to modulate cognitive functions in children. *BMC Pediatrics, 19*(1), 271. https://doi.org/10.1186/s12887-019-1637-x.

106. Yaribeygi, H., Panahi, Y., Sahraei, H., Johnston, T. P., & Sahebkar, A. (2017). The impact of stress on body function: A review. *EXCLI Journal, 16*, Doc1057. https://doi.org/10.17179/EXCLI2017-480.

107. Zhang, C., & Dong, C. (2023). The influence of social support on the mental health of elderly individuals in healthy communities with the framework of mental toughness. *Psychology Research and Behavior Management, 16*, 2977–2988. https://doi.org/10.2147/prbm.s413901.

INDEX

D

E

R

S

T

V

W